Contents

Dedications and acknwledgements

I should like to dedicate this book to my late father, grandmothers, grandfathers and uncles Petr and Ivan.

I should also like to express my special thanks to my mother.

Introduction

Perhaps, the main drawback of modern Algebraic Logic, going back to meaningless elaboration of so-called *abstract* (to be called more properly *orthodox* but sometimes ambitiously called even *universal*) *algebraic logic* (the principal activity of quite dishonest and perfectly untalented representatives of which consists in rewriting the genuine history of science for their own benefit), is that various associations of logical systems with algebras are often proposed without the least care of any practical usefulness of found connections for studying either particular logical systems with using advanced algebraic tools or, conversely, their algebraic counterparts upon the basis of constructive, proof-theoretical methods. This could not but give rise to the situation, when orthodox algebraists (especially, those around the algebraic journal, ambitiously called "Algebra Universalis", to be called more properly "Algebra Orthodoxis") frankly disregard constructive argumentation, while, conversely, orthodox logicians (in particular, those around the logic journal, ambitiously called "Logica Universalis", to be called more properly "Logica Orthodoxis") fail to comprehend algebraic argumentation. Such an unpleasant situation does nowise favour further development of Algebraic Logic.

For this reason, in our developing General Algebraic Logic (foundations of which are due to [42] and go back to [27]) we lay a special emphasis onto practical usefulness of discovered connections between logical systems and their algebraic counterparts in studying particular systems. The work [44] has been devoted to studying extensions of equivalential universal Horn theories through subprevarieties of their algebraic counterparts being prevarieties of algebraic systems (in Mal'cev's sense [19]).[1] The mentioned monograph demonstrates applicability of advanced tools of Universal Algebra (in Mal'cev's extended comprehension [19]) to exploring logical calculi of any type (viz., UHTs).

The aim of the present book is quite opposite — to demonstrate applications

---

[1]This issue goes back to [27] dealing with solely finitary UHTs, purely relational translations, algebraizable UHTs and, respectively, quasivarieties of pure algebras.

of advanced methods of Proof Theory to investigating classes of algebras. More spcifically, we start from strengthening the conception of Deduction Theorem (DT) (in the general sense of Appendix B of [43]) by involving a proper generic formulation of Peirce Law (PL) [22] that highlights the crucial difference between classical and intuitionistic implications. It appears that a certain universal Deduction Theorem (DT) schema, found first in Theorem 4.2 of [28] for merely sentential languages and recently extended to arbitrary first-order ones in Theorem B.7 of [43], respects PL as well. We then apply this general result to uniform argumentation of both *restricted equational definability of principal congruences (REDPC* in the sense of [8])[2] and *implicativity* (in the sense of [41]) in varieties of algebras related to Multiple-Valued Logic. This has proved possible due to the fact that the latters are, in a sense, algebraic counterparts of DT (with PL, respectively), following from Theorem B.3 of [43] (that was originally proved just as an auxiliary characterization of UHTs with DT used for proving Theorem 1.18 ibid., being a culminating result of the cited monograph)[3] together with both certain generic results of Section 1.4 of [42] concerning pre-varieties of algebras and Theorem B.2 of [43], stating that equivalence between UHTs respects DT and extended here to DT with PL as well. In this connection, before pursuing our introductary discussion, we need to recall the history of the issue involved.

The conception of Deduction Theorem, being a basic one of Proof Theory, has been specified on a generic level in [43]. It was the main result of the mentioned work (together with Theorem B.2 therein stating preservation of DT under equivalence between finitary UHTs studied in [42] and [27])[4] that has forced us to treat DT schemes (within the context of *sentential,* viz., propositional *Hilbert-style,* calculi) as rather collections of binary secondary connectives than just a single binary (possibly, secondary) connective as it had been standard before (cf. [45]). On the other hand, earlier (in 1995) a universal DT schema was discovered for *enlargable* propositional *sequent* (viz., *Gentzen-style*) calculi with structural rules and presented later on in [28] (cf. Theorem 4.2 therein).[5] Hence, once sequent calculi associated with finitely-

---

[2]This property has a fundamental value within Universal Algebra that we refrain from clarifying in detail here.

[3]This displays once more unexpected ways of development of fundamental science (General Algebraic Logic, in our case).

[4]Though the mentioned theorem was proved therein for purely-relational translations alone, it has been extended to arbitrary ones in Appendix A of [44] (cf. Theorem A.6 therein).

[5]The proof of it, following the standard proof-theoretical method of constructive argumentation of DT for the classical propositional calculus, described, e.g., in [20], is presented in proving Theorem B.7 of [44]. The crucial value of our DT for Proof Theory of Multiple-Valued

valued logics with equality determinant according to [38] are enlargable and have all structural rules (and so satisfy the DT with PL involved), while many of them are finitely-algebraizable (the works [39] and [40] have suggested effective criteria of their algebraizability), the algebraic counterparts of the latter possess implicativity (in particular, REDPC), while the constructive proof of Theorem B.2 of [43] immediately yields an effective construction of implementing schemes. This covers many well-known interesting varieties: Boolean algebras, (bounded) distributive lattices, (bounded) distributive bilattices [with negation], Kleene algebras [lattices], implicative De Morgan algebras [lattices], Łukasiewicz's finitely-valued algebras. In this way, we get a uniform constructive insight into the issues of implicativity of (in particular, REDPC in) such varieties, for it appears that these are immediate consequences of the generic DT given by Theorem 4.2 of [28] and PL proved here.

However, there are several quite interesting exceptions not equivalent to calculi of such a kind: namely, [Boolean] De Morgan algebras (lattices), Stone algebras. Nevertheless, these are still equivalent to appropriate extensions of certain calculi proposed in [38] by a common Weak Cotraposition rule. On the other hand, according to a culminating result of the present book motivating writing it, such extensions inherit (under a proper generic modification) our primary DT schema for initial enlargable calculi given by [38] and, in most cases (but the one related to Stone algebras, as it is proved in Section 3.3), PL as well. Thus, our generic approach does cover the exceptions involved as well. And what is more, as it is demonstrated in Subsection 3.2.2, this is applicable to sentential logics for providing quite non-trivial DT shemes for them. As a matter of fact, it was the quite non-trivial open problem of finding a DT schema for $\mathbb{TSBB}_4$ that initiated research underlying this book, while the successive solution of this problem in Subsection 3.2.2 demonstrates the entire success of the present study. In view of Boolean De Morgan algebras [33] being relationally equivalent purely-algebraic semantics for $\mathbb{TSBB}_4$, the main result of [46] providing REDPC for De Morgan algebras inherited by Boolean De Morgan algebras as well (in view of the argumentation of Lemma 3.1 of [44]), together with Theorems B.2 and B.3 of [43], we had already known that the logic under consideration had DT. However, we have refrained from any vain attempts *to guess right* the schema to be found but have concentrated on developing advanced generic tools, based upon our DT with PL for enlargable sequent calculi with structural rules discovered in Theorem 4.2 of [28], to deduce constructively

---

Logic has already been realized directly in [28] and has successfully been used further in [38] and, on proper extension to many-place sequent calculi, in [34], [37], [40].

the DT schema required. The fact that Boolean De Morgan algabras appeared among exceptions mentioned above has inevitably forced us to explore the issue of DT for extensions of primary contraposable enlargable multiple-conclusion sequent calculi by the Weak Contraposition rule. It then appeared that the elaborated generic tools are applicable not merely to the logic under consideration but equally to certain interesting varieties providing these with quite transparent and constructive deduction of implicativity (or, at least, REDPC). This completely clarifies the genuine origin of the study presented in this monograph. And what is more, this once more demonstrates unexpected ways of development of fundamental science, when certain generic tools, being elaborated for solving one task, become then applicable to solving another problems. As a matter of fact, it is such unexpected ways that justify the fundamental science that, in its turn, aims not solely in solving particular practical tasks but mainly in developing itself. Recall that, without following this basic paradigm, the humanity would ever assimilate neither electricity nor nuclear energy, while any negative aspects of such assimilation are definitely because of rather the nature of humans alone, being far from being perfect, than the fundamental science as such.

In this connection, we should like to highlight here that, though some of the studied above varieties have already been known to have REDPC, the purely algebraic argumentation of such results is normally too cumbersome and smooth, while schemes implementing REDPC can often hardly be extracted from proofs (see, e.g., [46], [14], [17] in this connection) because of the following oddity typical of the cited works: all they prove rather conditions equivalent (under the obvious conditions $\theta(a, b) = \theta(a \wedge b, a \vee b)$ and $a \wedge b \leqslant a \vee b$, valid for lattices) to REDPC than REDPC as such (for getting which one should just apply the substitution $[a/a \wedge b, b/a \vee b]$ to equations found therein).[6] This goes without saying that, following this way, one must first guess right schemes implementing REDPC, so genesis of such schemes eventually remained unexplained, while, following our constructive argumentation, all necessary schemes have quite naturally arisen via a proper combination of certain analytical expressions found in Appendix B of [43] (cf. Section 2.4 below) and Section 2.6 below. Thus, our study has, at least, methodological value in those cases, providing an essentially new (constructive) insight into those results. And what is more, the REDPC schemes found in the cited works are quite easily shown to be equivalent to ours.

As a supplementary material concerning PL, we provide a model-theoretic

_____

[6]In section 2.3 we briefly explain reasons of these drawbacks of purely algebraic proofs of such results.

characterization of finitary UHTs having DT with PL. We then apply it to extending Theorem 3.5 of [32] suggesting a semantics of enlargable multiple-conclusion sequent calculi with structural rules from sentential basic languages to arbitrary first-order ones. As a consequence, we provide new transparent proofs to the strong completeness results concerining those particular sequent calculi associated with finitely-valued logics with equality determinant according to [38] which are considered here. Though the argumentation found here (as opposed to the generic one provided ibid.) is rather *ad hoc*, it does yield the intuitive essence of the results involved. Besides that, it is based upon certain auxiliary lemmas equally providing transparent uniform proofs to the generation results for varieties under consideration.

The rest of the monograph is as follows. Chapter 1 is a concise self-contained summary of general set-theoretical and algebraic background adopted in [42] and needed for comprehending the present book in its own right. In Section 1.1 we make explicit an immediate consequence of a quite useful characterization of equivalent UHTs obtined in [42] (cf. Proposition 2.4 therein) to be used in Chapter 3 as well as, potentially, within another context. In Section 1.2 we analyze the issue of equivalence of translations to be equally used both in Chapter 3 and, potentially, elsewhere. Section 1.3 is an exposition of necessary results of Section 1.4 of [42] that are of crucial importance for the current study. Chapter 2 is devoted to elaborating certain general issues concerning both Deduction Theorem and Peirce Law. First of all, in Section 2.1 we formulate the conception of Peirce Law within the general context of UHTs and obtain a model-theoretic characterization of finitary UHTs having DT with PL. Section 2.2 extends Theorem A.6 of [44], stating preservation DT under equivalence between finitary UHTs, to DT with PL. Next, Section 2.3 argues that REDPRC [implicativity] is, so to say, an algebraic counterpart of DT [resp.,with PL]. Section 2.4 is the "PL-strengthening" of DT for enlargable multiple-conclusion sequent calculi discovered in [28]. In Section 2.5 we extend Theorem 3.5 of [32] concerning semantics of enlargable multi-conclusion sequent calculi with structural rules from sentential basic languages to arbitrary first-order ones. In Section 2.6 we adapt the DT schema, proposed in Section 2.4, for extensions of some of enlargable sequent calculi (called here *contraposable*) by the Weak Contraposition rule. Therein, we also study how such adaptation respects PL. Chapter 3 exemplifies our general elaboration by applying it to numerous examples of algebraizable calculi provided by the generic approach of [38] (roots of which go back to [23], [24], [26] and [28]) as well as to algebraizable extensions of those of them, which are contraposable but are not algebraizable themselves, by the Weak Contraposition rule. After all, Conclusions provide a concise purely-descriptive summary of principal contributions of the book. Finally, Appendix A is a brief summary of the generic approach of [38], further advanced within General Algebraic Logic in [39] (cf. [40]).

The exposition of the material is chosen to be almost self-contained. Unless otherwise specified, we entirely follow the customary conventions adopted in [27], [42] and [43]. In particular, when dealing with Deduction Theorem and sequent calculi, we tacitly follow [43]. Likewise, concerning equivalence and algebraizability, we tacitly follow both [42] and [44]. (All these advanced issues are too complicated and cumbersome to repeat them elsewhere.) For general basic issues concerning Lattice Theory, Universal Algebra and Model Theory, the reader is also referred to standard mathematical handbooks like [2], [4], [5], [11], [12], [13] and, especially, [19].

The present monograph incorporates those results of mine, which were obtained during my research starting from my post-graduated study since 1991 and which have proved beyond the scopes of my previous books [42], [43] and [44].

Chapter 1

General background

We follow the conventions adopted in [42]. Unless otherwise specified, we fix an arbitrary first-order signature $L = \langle F, R \rangle$, identified with F alone, whenever $R = \varnothing$. The arity of any symbol $s \in F \cup R$ in L is denoted by $\mu_L(s)$. For unifying notations, L-structures are denoted by Calligraphic letters $\mathcal{A}, \mathcal{B}, \mathcal{C}, \ldots$ (possibly, with indices), while [their underlying] F-algebras are denoted by [corresponding] Fraktur letters $\mathfrak{A}, \mathfrak{B}, \mathfrak{C}, \ldots$ (with [the same] indices if any), whereas their carriers are denoted by respective capital Italic letters $A, B, C, \ldots$ (with respective indices if any). The class of all F-algebras [L-structures] is denoted by A_F [resp., by S_L].

Natural numbers (including 0) are treated as ordinals, i.e., sets constituted by lesser natural numbers, the countable ordinal of all them being denoted by ω. The class of all ordinals is denoted by ∞. (The expression $\alpha < \infty$ means $\alpha \in \infty$.) During the book, κ is supposed to be an arbitrary regular cardinal (in particular, ω).

Script letters x, y, z (possibly, with indices) are used for denoting variables. Given any $\alpha \leqslant \infty$, put $V^\alpha \triangleq \{x_\beta\}_{\beta \in \alpha}$. The class of all F-terms (equality-free atomic L-formulas) with variables in V^α is denoted by $\mathrm{Tm}_F(V^\alpha)$ (resp., by $\mathrm{At}_L(V^\alpha)$). The set of all variables really occurring in any $\Phi \in \mathrm{Tm}_F(V^\infty) \cup \mathrm{At}_L(V^\infty)$ is denoted by $\mathrm{Var}(\Phi)$. Given any $n \in \omega$ and $\alpha \in \infty$, set $\bar{x}(\alpha, n) \triangleq (x_{\alpha+i})_{i \in n} \in (V^\infty)^n$.

As usual, singletons are naturally identified with their unique elements, unless any confusion is possible (like in case of ordinals, for, e.g., $1 = \{0\} \neq 0$).

Let A be a set. As usual, put $A^* \triangleq \bigcup_{n \in \omega} A^n$. The *power set of A*, constituted by all subsets of A, is denoted by $\wp(A)$. Given any $B \subseteq A$, the *characteristic function χ_A^B of $B \subseteq A$ in A* is the mapping from A to 2, given by $\chi_A^B[B] = \{1\}$

and $\chi_A^B[A \setminus B] = \{0\}$. This provides the standard identification of $\wp(A)$ with 2^A. An *equivalence on A* is any reflexive, transitive and symmetric binary relation E on A, in which case we have the *natural mapping* $\nu_E : A \to A/E \triangleq \{[a]_E | a \in A\}, a \mapsto [a]_E \triangleq \{b \in A | aEb\}$. An example of an equivalence on A is the *diagonal relation* (viz., *identity mapping*) $\Delta_A \triangleq \{\langle a, a \rangle | a \in A\}$ *on A*. In addition, given any function h with $\operatorname{dom} h = A$, its *kernel* $\ker h \triangleq \{\langle a, b \rangle \in A^2 | ha = hb\} = h^{-1}[\Delta_{\operatorname{img} h}]$ is an equivalence on A. Moreover, $\ker \nu_E = E$.

Let $\langle P, \leqslant \rangle$ be a poset forming a complete lattice with infinitary join $\bigvee$ and meet $\bigwedge$ operations. A *closure system on P* is any $\mathcal{C}$ forming a (complete) $\bigwedge$-subsemilattice of P. An $a \in P$ is said to be an *upper bound of* an $S \subseteq P$, provided $b \leqslant a$, for each $b \in S$. A $c \in S$ is said to be *maximal*, provided every upper bound of c belonging to S is equal to c itself. A $U \subseteq P$ is said to be *upward-directed*, provided every finite $S \subseteq U$ has an upper bound belonging to U (in which case $U \neq \varnothing$, when taking $S = \varnothing$). An $S \subseteq P$ is said to be *inductive*, whenever, for each upward-directed $U \subseteq S$, it holds that $\bigvee U \in S$. In that case, by contradiction, using Choice Axiom, by constructing (via transfinite induction) an embedding of $\langle \infty, \subset \rangle$ into $\langle S, < \rangle$, it is proved that S has a maximal element, whenever it is not empty. A *closure operator on P* is any $C : P \to P$ such that, for all $a, b \in P$ such that $a \leqslant b$, it holds that $\bigvee\{C(C(a)), a\} \leqslant C(a) \leqslant C(b)$, in which case $\operatorname{img} C$ is a closure system on P said to be *dual to C*. Given a set A, a *closure system (operator) over A* is any closure system (operator) on $\langle \wp(A), \subseteq \rangle$, in which case $\bigvee = \bigcup$, while $\bigwedge = \bigcap$.

Let $\mathfrak{A}$ be an *F-algebra*. A *congruence of $\mathfrak{A}$* is any equivalence θ on A such that, for each $f \in F$ and all $\bar{a}, \bar{b} \in A^{\mu_L(f)}$ such that $a_i \theta b_i$, for every $i \in \mu_L(f)$, it holds that $f^{\mathfrak{A}}(\bar{a}) \theta f^{\mathfrak{A}}(\bar{b})$, the set of all them being denoted by $\operatorname{Con}(\mathfrak{A})$. Given any function h with $\operatorname{dom} h = A$ and $\ker h \in \operatorname{Con}(\mathfrak{A})$, we have the *image* $h[\mathfrak{A}] \triangleq \langle \operatorname{img} h, h \circ f^{\mathfrak{A}} \circ h^{-1} \rangle_{f \in F}$ of $\mathfrak{A}$ by h. Given any $\theta \in \operatorname{Con}(\mathfrak{A})$, $\mathfrak{A}/\theta \triangleq \nu_\theta[\mathfrak{A}]$ is referred to as the *quotient of $\mathfrak{A}$ by θ*. Given a one more *F-algebra $\mathfrak{B}$*, a *homomorphism from $\mathfrak{A}$ to $\mathfrak{B}$* is any $h : A \to B$ such that, for each $f \in F$ and every $\bar{a} \in A^{\mu_L(f)}$, it holds that $hf^{\mathfrak{A}}(\bar{a}) = f^{\mathfrak{B}}(h\bar{a})$, in which case $\ker h \in \operatorname{Con}(\mathfrak{A})$, the set of all them being denoted by $\operatorname{Hom}(\mathfrak{A}, \mathfrak{B})$. Note that, for any $\theta \in \operatorname{Con}(\mathfrak{A})$, $\nu_\theta \in \operatorname{Hom}(\mathfrak{A}, \mathfrak{A}/\theta)$.

Let $\mathcal{A}$ be an *L-structure*. A *congruence of $\mathcal{A}$* is any $\theta \in \operatorname{Con}(\mathfrak{A})$ such that, for each $r \in R$ and all $\bar{a}, \bar{b} \in A^{\mu_L(r)}$ such that $a_i \theta b_i$, for every $i \in \mu_L(r)$, it holds that $\bar{a} \in r^{\mathfrak{A}} \Leftrightarrow \bar{b} \in r^{\mathfrak{A}}$, the set of all them being denoted by $\operatorname{Con}(\mathcal{A})$. According to Proposition 1.4 of [42], one can equivalently set $\mathfrak{S}(\mathcal{A}) \triangleq \bigcup \operatorname{Con}(\mathcal{A}) \in \operatorname{Con}(\mathcal{A})$. Then, $\mathcal{A}$ is said to be *simple*, if $\mathfrak{S}(\mathcal{A}) = \Delta_A$. Given any function h with $\operatorname{dom} h = A$ and $\ker h \in \operatorname{Con}(\mathfrak{A})$, we have the *image* $h[\mathcal{A}] \triangleq \langle h[\mathfrak{A}], h[r^A] \rangle_{r \in R}$ of $\mathcal{A}$ by h. Given any $\theta \in \operatorname{Con}(\mathfrak{A})$, $\mathcal{A}/\theta \triangleq \nu_\theta[\mathcal{A}]$ is referred to as the *quotient of A by θ*. Given a one more L-structure $\mathcal{B}$, a *(strict) homomorphism from A to B* is any $h \in \mathrm{Hom}(\mathfrak{A}, \mathfrak{B})$ such that, for each $r \in R$, $r^{\mathcal{A}} \subseteq (=)h^{-1}[r^{\mathcal{B}}]$ (in which case $\ker h \in \mathrm{Con}(\mathcal{A})$), the set of all them being denoted by $\mathrm{Hom}(\mathcal{A}, \mathcal{B})$ (resp., by $\mathrm{SHom}(\mathcal{A}, \mathcal{B})$). Note that, for any $\theta \in \mathrm{Con}(\mathfrak{A})(\in \mathrm{Con}(\mathcal{A}))$, $\nu_\theta \in \mathrm{Hom}(\mathcal{A}, \mathcal{A}/\theta)$ (resp., $\nu_\theta \in \mathrm{SHom}(\mathcal{A}, \mathcal{A}/\theta)$). Injective (and surjective) strict homomorphims are referred to as *embeddings* (resp., *isomorphisms*). Given an F-algebra $\mathfrak{A}$, an L-structure $\mathcal{B}$ and any $h \in \mathrm{Hom}(\mathfrak{A}, \mathfrak{B})$, we have the L-structure $h^{-1}[\mathcal{B}] \triangleq \langle \mathfrak{A}, h^{-1}[r^{\mathcal{B}}]\rangle_{r \in R}$. In case $h = \Delta_A$, we set $\mathcal{B} \upharpoonright A \triangleq h^{-1}[\mathcal{B}]$. Structures of such a form are referred to as *substructures of* $\mathcal{B}$, the set of all them being denoted by $\mathbf{S}(\mathcal{B})$.

An L-structure $\mathcal{A}$ is said to be *total*, whenever $r^{\mathfrak{A}} = A^{\mu_L(r)}$, for every $r \in R$, and *proper*, otherwise.

Put $(L + \approx) \triangleq \langle F, R + \approx \rangle$, where $(R + \approx) \triangleq R \cup \{\approx\}$, while $\mu_{L+\approx}$ is the extension of μ_L to $F \cup (R + \approx)$ defined by $\mu_{L+\approx}(\approx) \triangleq 2$. Any $(L + \approx)$-structure $\mathcal{A}$ is naturally identified with the couple $\langle \mathcal{A} \upharpoonright L, \approx^{\mathcal{A}} \rangle$. Then, given an L-structure $\mathcal{A}$, we have the $(L + \approx)$-structure $(\mathcal{A} + \approx) \triangleq \langle \mathcal{A}, \Delta_A \rangle$. Given a class of L-structures K, set $(\mathsf{K} + \approx) \triangleq \{\mathcal{A} + \approx | \mathcal{A} \in \mathsf{K}\}$. Recall that $\vdash_\mathsf{K}$, called *the consequence of* K, denotes the class of all *equality-free* infinitary universal Horn L-sentences[1] (viz. *-rules* or *-implications*) true in each member of K, in which case:

(1.1) $$\vdash_\mathsf{K} = \vdash_{\mathsf{K}'},$$

where K' is the class of all proper members of K. In this way, $\vdash_\mathsf{K}^{\approx} \triangleq \vdash_{\mathsf{K}+\approx}$ does actually involve sentences with equality. Given an equality-free (κ-ary) universal Horn L-theory $\mathbb{T}$ (that is, such that every $(\Gamma \to \Phi) \in \mathbb{T}$ is κ-ary, i.e., $|\Gamma| < \kappa$).[2] By $\mathrm{Mod}(\mathbb{T})$ we denote the class of all *models of* $\mathbb{T}$, that is, those L-structures, which satisfy all sentences in $\mathbb{T}$, in which case, according to Theorem 1.6 of [42], one can equivalently set $\vdash_\mathbb{T} \triangleq \vdash_{\mathrm{Mod}(\mathbb{T})}$, rules in which are said to be *derivable in* $\mathbb{T}$, according to Appendix A of [43], while $\vdash_\mathbb{T}$ is said to be *axiomatized by* $\mathbb{T}$. Then, an *extension of* $\mathbb{T}$ is any $\vdash_{\mathbb{T}'} \supseteq \vdash_\mathbb{T}$, said to be *relatively axiomatized by* $\mathbb{T}' \setminus \mathbb{T}$. By $\mathrm{Mod}_{\mathfrak{S}}(\mathbb{T})$ we denote the class of all simple models of $\mathbb{T}$. *L-Axioms* (viz., *L-identities*) are premise-less L-rules. Then, an extension is said to be *axiomatic*, if it is relatively axiomatized by axioms alone. Next, given any (κ-ary) universal Horn $(L + \approx)$-theory $\mathbb{T}$, put $\mathrm{Mod}_{\approx}(\mathbb{T}) \triangleq \{\mathcal{A} \in \mathbf{S}_L | (\mathcal{A} + \approx) \in \mathrm{Mod}(\mathbb{T})\}$ (this is the class of all models of $\mathbb{T}$ in the standard comprehension of Model Theory with equality). Then,

---

[1] As usual, we omit universal prefixes in writing any universal sentences.

[2] As usual, we say "finitary" for "ω-ary" within any context.

a *(κ-ary) prevariety* (viz., *implicational class*) is any class of such a kind, in which case it is said to be *axiomatized by* $\mathbb{T}$. (Finitary prevarieties are referred to as *quasivarieties*, while finitary rules are also called *quasi-identities*.) A *variety* is any prevariety axiomatized by an UHT consisting of identities alone. The prevariety ([quasi]variety) *generated by* $\mathsf{K} \subseteq \mathsf{S}_L$ is defined to be the one axiomatized by all implications (resp. [quasi-]identities) in $\vdash_{\mathsf{K}}^{\approx}$. (It is the least one including K.) This is denoted by $\mathbf{I}(\mathsf{K})$ (resp., by $[\mathbf{Q}(\mathsf{K})]$ $\mathbf{V}(\mathsf{K})$). According to [19], an L-structure $\mathcal{A} \in \mathbf{I}(\mathsf{K})$ iff $\mathcal{A}$ is isomorphic to a substructure of a direct product of members of K or, equivalently, $\bigcap\{\ker h | h \in \mathrm{Hom}(\mathcal{A}, \mathcal{B}), \mathcal{B} \in \mathsf{K}\} = \Delta_A$ and $\bigcap\{h^{-1}[r^{\mathcal{B}}] | h \in \mathrm{Hom}(\mathcal{A}, \mathcal{B}), \mathcal{B} \in \mathsf{K}\} = r^{\mathcal{A}}$, for each $r \in R$.

An *inverse rule of* an L-rule $\Gamma \rightarrow \Phi$ is any L-rule of the form $\{\Phi\} \rightarrow \Psi$, where $\Psi \in \Gamma$.

As usual, when dealing with sequent languages, we use fraction notations for rules.

Let $\mathfrak{A}$ be an F-algebra. The set of L-structures with underlying algebra $\mathfrak{A}$ is denoted by $\mathrm{S}_L(\mathfrak{A})$. Put $\mathrm{At}_L(\mathfrak{A}) \triangleq \{r(\bar{a}) | r \in R, \bar{a} \in A^{\mu_L(r)}\}$. Given any $\mathcal{A} \in \mathrm{S}_L(\mathfrak{A})$, we have $(\mathcal{A} \downarrow) \triangleq \{r(\bar{a}) | r \in R, \bar{a} \in r^{\mathcal{A}}\} \subseteq \mathrm{At}_L(\mathfrak{A})$ (given any $\Phi \in \mathrm{At}_L(\mathfrak{A})$, we naturally write $\mathcal{A} \models \Phi$ for $\Phi \in (\mathcal{A} \downarrow)$). Conversely, given any $\Gamma \subseteq \mathrm{At}_L(\mathfrak{A})$, we have the $(\Gamma \uparrow \mathfrak{A}) \in \mathrm{S}_L(\mathfrak{A})$, defined by $r^{(\Gamma \uparrow \mathfrak{A})} \triangleq \{\bar{a} | r(\bar{a}) \in \Gamma\}$, for every $r \in R$. These yield inverse to one another bijections between $\mathrm{S}_L(\mathfrak{A})$ and $\wp(\mathrm{At}_L(\mathfrak{A}))$, in which case the partial ordering on the former induced by the one $\subseteq$ on the latter is denoted by $\preceq$, the induced meet and join operations retaining the primary notations $\bigcap$ and $\bigcup$, respectively. Given any universal L-theory $\mathbb{T}$, the set $\mathrm{M}_{\mathbb{T}}(\mathfrak{A}) \triangleq \mathrm{S}_L(\mathfrak{A}) \cap \mathrm{Mod}(\mathbb{T})$ is a closure system on $\langle \mathrm{S}_L(\mathfrak{A}), \preceq \rangle$, the dual closure operator being denoted by $\mathrm{Lm}_{\mathbb{T}}^{\mathfrak{A}}$.

1.1 Relative axiomatizations of extensions of equivalent universal Horn Theories

Here, we entirely follow Chapter 2 of [42].

Corollary 1.1. *Let* $\mathbb{T}_1$, $\mathbb{T}_2$, τ *and* ρ *be as in Definition 2.1 of [42], while, for each* $i \in 3 \setminus 1$, $\mathbb{T}_i'$ *is a universal Horn* L_i-*theory such that* $\mathbb{T}_i \subseteq \mathbb{T}_i'$. *Suppose* $\mathbb{T}_1$ *and* $\mathbb{T}_2$ *are equivalent with respect to* τ *and* ρ. *Then,* $\mathbb{T}_1'$ *and* $\mathbb{T}_2'$ *are equivalent with respect to* τ *and* ρ *iff the following two conditions hold:*

(i) $\tau[\mathbb{T}_1' \setminus \mathbb{T}_1] \subseteq \vdash_2$;

(ii) $\rho[\mathbb{T}_2' \setminus \mathbb{T}_2] \subseteq \vdash_1$.

13

Proof. By Proposition 2.4 of [42], for the conditions 2.1(iii-iv) and 2.7(v,vi) therein being valid for $\mathbb{T}_1$ and $\mathbb{T}_2$ remain true for $\mathbb{T}'_1$ and $\mathbb{T}'_2$ as well. $\qquad\square$

1.2   Equivalent translations

Here, we equally follow Chapter 2 of [42].

Two (L_1, L_2)-translations τ and τ' are said to be *equivalent with respect to* a universal Horn L_2-theory $\mathbb{T}$ ($\tau \cong_{\mathbb{T}} \tau'$, in symbols), provided $\tau(f) \equiv_{\mathbb{T}} \tau'(f)$, for every $f \in F_1$, while $\tau(r) \dashv\vdash_{\mathbb{T}} \tau'(r)$, for every $r \in R_1$.

Lemma 1.2. *Let $\mathbb{T}$, τ and τ' be as above. Suppose $\tau \cong_{\mathbb{T}} \tau'$. Then, the following hold:*

(i) $\tau(\phi) \equiv_{\mathbb{T}} \tau'(\phi)$, *for every $\phi \in \mathrm{Tm}_1(V^\infty)$;*

(ii) $\tau(\Phi) \dashv\vdash_{\mathbb{T}} \tau'(\Phi)$, *for every $\Phi \in \mathrm{At}_1(V^\infty)$.*

Proof. (i) is proved by induction on construction of ϕ. First, assume $\phi \in V^\infty$. Then, $\tau\phi = \phi = \tau'\phi$, so $\tau(\phi) \equiv_{\mathbb{T}} \tau'(\phi)$, by reflexivity of $\equiv_{\mathbb{T}}$. Next, assume $\phi = f(\overline{\phi})$, where $f \in F_1$ of arity $n \in \omega$, while $\overline{\phi} \in \mathrm{Tm}_1(V^\infty)^n$. Suppose $\tau(\phi_i) \equiv_{\mathbb{T}} \tau'(\phi_i)$, for each $i \in n$. Then, we have:

$$
\begin{aligned}
\tau(\phi) = \ & \tau(f)[x_i/\tau(\phi_i)] \\
\equiv_{\mathbb{T}} \ & \tau'(f)[x_i/\tau(\phi_i)] \quad \text{by } \tau \cong_{\mathbb{T}} \tau' \text{ and the full invariance of } \equiv_{\mathbb{T}} \\
& \qquad\qquad\qquad\quad (\text{cf. Proposition 1.11 of [42]}) \\
\equiv_{\mathbb{T}} \ & \tau'(f)[x_i/\tau'(\phi_i)] \quad \text{by Corollary 1.8 of [42] with} \\
& \qquad\qquad\qquad\quad V = V^n \text{ and } V' = \mathrm{Var}(\phi) \cup V^1 \\
= \ & \tau'(\phi).
\end{aligned}
$$

This completes the argument of (i).

For proving (ii), take any $\Phi \in \mathrm{At}_1(V^\infty)$. Then, $\Phi = r(\overline{\phi})$, where $r \in R_1$ of arity $n \in \omega$, while $\overline{\phi} \in \mathrm{Tm}_1(V^\infty)^n$, in which case, by (i), we have $\tau(\phi_i) \equiv_{\mathbb{T}} \tau'(\phi_i)$, for each $i \in n$. Then, we get:

$$
\begin{aligned}
\tau(\Phi) = \ & \tau(r)[x_i/\tau(\phi_i)] \\
\dashv\vdash_{\mathbb{T}} \ & \tau'(r)[x_i/\tau(\phi_i)] \quad \text{by } \tau \cong_{\mathbb{T}} \tau' \text{ and the structurality of } \vdash_{\mathbb{T}} \\
\dashv\vdash_{\mathbb{T}} \ & \tau'(r)[x_i/\tau'(\phi_i)] \quad \text{by Corollary 1.10 of [42] with} \\
& \qquad\qquad\qquad\quad V = V^n \text{ and } V' = \mathrm{Var}(\Phi) \cup V^1 \\
= \ & \tau'(\Phi).
\end{aligned}
$$

This completes the argument of (ii). $\qquad\square$

Proposition 1.3. *Let $\mathbb{T}_i$ be a universal Horn L_i-theory, where $i \in 3 \setminus 1$, τ and τ' (L_1, L_2)-translations equivalent with respect to $\mathbb{T}_2$, while ρ and ρ' (L_2, L_1)-translations equivalent with respect to $\mathbb{T}_1$. Then, $\mathbb{T}_1$ and $\mathbb{T}_2$ are equivalent with respect to τ and ρ iff they are equivalent with respect to τ' and ρ'.*

Proof. Straightforward, with using Lemma 1.2(ii) and the structurality of both $\vdash_1$ and $\vdash_2$ (to be involved in checking 2.1(v,vi) of [42]). $\qquad\square$

1.3 Universal Horn theories of prevarieties of algebras

As usual, given any $\alpha \in \infty$, elements of $\mathrm{Eq}_F(V^\alpha) \triangleq \mathrm{At}_{F+\approx}(V^\alpha)$ are referred to as *F-equations over V^α*.

Remark 1.4. In view of Proposition 1.17 of [44], for any F-algebra $\mathfrak{A}$, the $(F+\approx)$-structure $\mathfrak{A} + \approx$ is simple. $\qquad\square$

Given a class of F-algebras K and an F-algebra $\mathfrak{A}$, a K-*(relative)congruence of* $\mathfrak{A}$ is any $\theta \in \mathrm{Con}(\mathfrak{A})$ such that $\mathcal{A}/\theta \in$ K, the set of all them being denoted by $\mathrm{Con}_{\mathsf{K}}(\mathfrak{A})$. Then, $\mathcal{A}$ is said to be *algebraically* K-*(relatively)simple*, whenever $\mathrm{Con}_{\mathsf{K}}(\mathfrak{A}) \subseteq \{\Delta_A, A^2\}$.[3] If K is an infravariety (in particular, prevariety), in view of Lemma A.1 of [42], $\mathrm{Con}_{\mathsf{K}}(\mathfrak{A})$ is a closure system over A^2, in which case $\mathrm{Cg}_{\mathsf{K}}^{\mathfrak{A}}$ denotes the dual closure operator. (Note that $\mathrm{Con}_{\mathsf{K}}(\mathfrak{A}) = \mathrm{Con}(\mathfrak{A})$, whenever K is a variety, while $\mathfrak{A} \in$ K, in which case any mention of K is omitted.)

Recall that, by (1.14), (1.15) and Corollary 1.24 of [42], we have:

$$(1.2) \qquad \mathrm{Mod}_3(\vdash_{\mathsf{K}}^{\approx}) = (\mathsf{K} + \approx),$$

provided K is a prevariety. This enables us to prove:

Proposition 1.5. *Let* K *be a prevariety of F-algebras. Then,*

$$\mathrm{Mod}(\vdash_{\mathsf{K}}^{\approx}) = \{\langle \mathfrak{A}, \theta \rangle | \mathfrak{A} \in \mathsf{A}_F, \theta \in \mathrm{Con}_{\mathsf{K}}(\mathfrak{A})\}.$$

---

[3]The reservation "algebraically", never involved within Universal Algebra, proves to be well-justified within General Algebraic Logic to avoid any confusion with the notion of simple algebraic system (cf. [42]). The thing is that, as opposed to pure algebras, the totality binary relation on the carrier of an algebraic system is normally not a congruence of the latter, unless this is total (such is obviously the case for pure algebras). On the other hand, it is proper algebraic systems that deserve to be studied within Universal Horn Logic without equality (viz., General Propositional Logic), in view of (1.1).

15

Proof. First, take any $\mathcal{A} \in \mathrm{Mod}(\vdash_{\mathsf{K}}^{\approx})$. Then, by (1.14) and Proposition 1.23 of [42], we have $\mathcal{A} = \langle \mathfrak{A}, \Im(\mathcal{A}) \rangle$. Moreover, by Lemma 1.14 and Corollary 1.5 of [42], $\mathcal{A}/\Im(\mathcal{A}) \in \mathrm{Mod}_{\Im}(\vdash_{\mathsf{K}}^{\approx})$. On the other hand, $\mathcal{A}/\Im(\mathcal{A}) = ((\mathfrak{A}/\Im(\mathcal{A})) + \approx)$, so, by (1.2), $\Im(\mathcal{A}) \in \mathrm{Con}_{\mathsf{K}}(\mathfrak{A})$, as required. Conversely, consider any F-algebra $\mathfrak{A}$ and any $\theta \in \mathrm{Con}_{\mathsf{K}}(\mathfrak{A})$. Then, by (1.2), $\mathcal{B} \triangleq ((\mathfrak{A}/\theta) + \approx) \in \mathrm{Mod}(\vdash_{\mathsf{K}}^{\approx})$. Therefore, in view of Lemma 1.14 of [42], $\langle \mathfrak{A}, \theta \rangle = (\nu_{\theta})^{-1}[\mathcal{B}] \in \mathrm{Mod}(\vdash_{\mathsf{K}}^{\approx})$ as well. This completes the argument. $\qquad\square$

Chapter 2

Deduction Theorem and Peirce Law

As preliminary generic points, we start from presenting the following useful auxiliary observations supplementing those presented in Appendix B of [43] and being immediate consequences of, resp., (B.2) and (B.7) ibid.:

Proposition 2.1. *Let $\mathbb{T}$ be a universal Horn L-theory, and both λ and λ' binary L-systems. Suppose $\mathbb{T}$ has DT with respect to λ. Then, $\mathbb{T}$ has DT with respect to λ' iff $\lambda \cong_{\mathbb{T}} \lambda'$.*

Proposition 2.2. *Let $\mathbb{T}$ be a universal Horn L-theory, λ a binary L-system, $m, n \in \omega$, $\alpha \in \infty$, $\Gamma \subseteq \mathrm{At}_L(V^\alpha)$, $\overline{\Phi} \in \mathrm{At}_L(V^\alpha)^m$ and $\overline{\Psi} \in \mathrm{At}_L(V^\alpha)^n$. Suppose $\mathbb{T}$ has DT with respect to λ, while $(\mathrm{img}\,\overline{\Phi}) \dashv\vdash_{\mathbb{T}} (\mathrm{img}\,\overline{\Psi})$ (in particular, $(\mathrm{img}\,\overline{\Phi}) = (\mathrm{img}\,\overline{\Psi})$). Then, $\lambda(\overline{\Phi}, \Gamma) \dashv\vdash_{\mathbb{T}} \lambda(\overline{\Psi}, \Gamma)$.*

2.1 Deduction Theorem with Peirce Law

Consider any universal Horn L-theory $\mathbb{T}$ having DT with respect to any binary L-system λ. Then, $\mathbb{T}$ is said to *hold Peirce Law (PL)* with respect to λ, provided:

$$(2.1) \qquad \lambda(\Omega, \Phi) \vdash_{\mathbb{T}} \Phi,$$

for all $\Phi, \Psi \in \mathrm{At}_L(V^\infty)$ and every $\Omega \in \lambda(\Phi, \Psi)$. In the sentential case, when, in addition, λ is a singleton (viz., a secondary binary connective), (2.1) is equivalent to the standard comprehension of Peirce Law (cf. [22]):

$$(2.2) \qquad \varnothing \vdash_{\mathbb{T}} \lambda(\lambda(\lambda(x_0, x_1), x_0), x_0).$$

Given a binary L-system λ, an L-structure $\mathcal{A}$ is said to be λ(-*implicative*) *[-prime]*, provided it is compatible with λ and, for all $\Phi, \Psi \in \mathrm{At}_L(\mathfrak{A})$:

$$(2.3) \qquad (\mathcal{A} \models \Psi)[\mathcal{A} \not\models \Phi] \Rightarrow \lambda^{-1}[\mathcal{A}] \models \langle \Phi, \Psi \rangle (\Rightarrow \mathcal{A} \models \Phi \Rightarrow \mathcal{A} \models \Psi).$$

Remark 2.3. Given any universal Horn L-theory $\mathbb{T}$ having DT with respect to λ, in view Theorem B.3(i)$\Rightarrow$(iv)**b),c),e)** of [43], any model of $\mathbb{T}$ is λ-implicative. $\qquad\square$

Theorem 2.4. *Let $\mathbb{T}$ be a finitary universal Horn L-theory and λ a binary L-system. Then, the following are equivalent:*

  (i) $\mathbb{T}$ *has DT and holds PL with respect to λ;*

  (ii) $\vdash_{\mathbb{T}}$ *is the consequence of the class of all λ-prime λ-implicative models of $\mathbb{T}$;*

  (iii) $\vdash_{\mathbb{T}}$ *is the consequence of the class of all simple λ-prime λ-implicative models of $\mathbb{T}$;*

  (iv) $\vdash_{\mathbb{T}}$ *is the consequence of a class of λ-prime λ-implicative L-structures;*

  (v) $\vdash_{\mathbb{T}}$ *is the consequence of a class of simple λ-prime λ-implicative L-structures.*

Proof. (i)$\Rightarrow$(iv).
Assume (i) holds. Then, by Theorem B.3 of [43], $\mathbb{T}$ has ADT with respect to λ over $\mathrm{Mod}(\mathbb{T})$, in which case the latter is compatible with λ. Consider any $\mathcal{A} \in \mathrm{Mod}(\mathbb{T})$. The set of all λ-prime members of $\mathrm{M}_{\mathbb{T}}(\mathfrak{A})$, in which case they are λ-implicative, in view of Remark 2.3, is denoted by $\mathrm{PM}_{\mathbb{T}}^\lambda(\mathfrak{A})$. Put:

$$\mathcal{G}_\mathcal{A} \triangleq \{ \mathcal{B} \in \mathrm{PM}_{\mathbb{T}}^\lambda(\mathfrak{A}) | \mathcal{A} \preceq \mathcal{B} \}.$$

Clearly, $\mathcal{A} \preceq \bigcap \mathcal{G}_\mathcal{A}$. For proving the converse, take any $\Upsilon \in \mathrm{At}_L(\mathfrak{A})$ such that $\mathcal{A} \not\models \Upsilon$. Consider the family $\mathcal{H}_\mathcal{A}^\Upsilon \triangleq \{ \mathcal{B} \in \mathrm{M}_{\mathbb{T}}(\mathfrak{A}) | \mathcal{A} \preceq \mathcal{B} \not\models \Upsilon \}$. Since $\mathbb{T}$ is finitary, $\mathrm{M}_{\mathbb{T}}(\mathfrak{A})$ is inductive, in which case $\mathcal{H}_\mathcal{A}^\Upsilon$ is inductive too. Moreover, it is non-empty, for it contains $\mathcal{A}$, so it has a maximal element $\mathcal{E}$. We claim that it is λ-prime. For take any $\Phi', \Psi' \in \mathrm{At}_L(\mathfrak{A})$. Assume $\mathcal{E} \not\models \Phi'$. Then, by the maximality of $\mathcal{E}$, we have:

$$(2.4) \qquad \mathrm{Lm}_{\mathbb{T}}^{\mathfrak{A}}((\mathcal{E} \downarrow \cup \{\Phi'\}) \uparrow \mathfrak{A}) \models \Upsilon.$$

Let us prove by contradiction that $\lambda^{-1}[\mathcal{E}] \models \langle \Upsilon, \Psi' \rangle$. For suppose $\lambda^{-1}[\mathcal{E}] \not\models \langle \Upsilon, \Psi' \rangle$. Assume $\Upsilon = r(\bar{c})$, while $\Psi' = p(\bar{b})$, for some $r, p \in R$. Then, there is

18

some $\Omega \triangleq q(\bar{\phi}) \in \lambda(r, p)$, where $q \in R$, such that $\mathcal{E} \not\models \Omega' \triangleq q(\bar{\phi}^{\mathfrak{A}}[h])$, where $h : V^{\mu_L(r) + \mu_L(p)} \cup V^1 \to A$, defined by:

$$
h x_i \triangleq \begin{cases} c_i & \text{if } i \in \mu_L(r), \\ b_{i - \mu_L(r)} & \text{if } i \in ((\mu_L(r) + \mu_L(p)) \setminus \mu_L(r)), \\ a & \text{elsewhere,} \end{cases}
$$

for all $i \in ((\mu_L(r) + \mu_L(p)) \cup 1$, where $a \in A$. Then, by the maximality of $\mathcal{E}$, we would have $\mathrm{Lm}_{\mathbb{T}}^{\mathfrak{A}}((\mathcal{E} \downarrow \cup \{\Omega'\}) \uparrow \mathfrak{A}) \models \Upsilon$, in which case, by (B.1) of [43], we would get $\lambda^{-1}[\mathcal{E}] \models \langle \Omega', \Upsilon \rangle$, and so by (2.1) with $\Phi \triangleq r(\bar{x}(0, \mu_L(r)))$ and $\Psi \triangleq p(\bar{x}(\mu_L(r), \mu_L(p)))$ under the assignment h and Note 2.3, we would have $\mathcal{E} \models \Upsilon$. By this contradiction, we conclude that $\lambda^{-1}[\mathcal{E}] \models \langle \Upsilon, \Psi' \rangle$. Then, by Note 2.3 and (2.4), we get $\mathrm{Lm}_{\mathbb{T}}^{\mathfrak{A}}((\mathcal{E} \downarrow \cup \{\Phi'\}) \uparrow \mathfrak{A}) \models \Psi'$, so, by (B.1) of [43], we obtain $\lambda^{-1}[\mathcal{E}] \models \langle \Phi', \Psi' \rangle$. Thus, $\mathcal{E}$ is indeed λ-prime. Therefore, $\mathcal{E} \in \mathcal{G}_A$, in which case $\bigcap \mathcal{G}_A \not\models \Upsilon$, for $\mathcal{E} \not\models \Upsilon$. Hence, $\mathcal{A} = \bigcap \mathcal{G}_A$, in which case we have:

$$
(2.5) \qquad\qquad \vdash_{\mathcal{G}_A} \subseteq \vdash_A.
$$

Put $\mathsf{K} \triangleq \bigcup \{\mathcal{G}_A | \mathcal{A} \in \mathrm{Mod}(\mathbb{T})\}$. Then, by Theorem 1.6 of [42] and (2.5), we eventually obtain $\vdash_{\mathbb{T}} = \vdash_{\mathsf{K}}$. Thus, (iv) does hold as required.

Next, (iv)$\Rightarrow$(ii) is by Theorem 1.6 of [42], while the converse is trivial. Further, (iv)$\Rightarrow$(i) is straightforward. Furthermore, (v)$\Rightarrow$(iii) is by Corollary 1.15 of [42], while the converse is trivial. Likewise, (v)$\Rightarrow$(iv) is trivial. Finally, for proving the converse, assume (iv) holds. Take any class K of λ-prime λ-implicative L-structures such that $\vdash_{\mathbb{T}} = \vdash_{\mathsf{K}}$. Put $\mathsf{K}' \triangleq \{\mathcal{A}/\mathfrak{S}(\mathcal{A}) | \mathcal{A} \in \mathsf{K}\}$. By Corollary 1.5 of [42], each member of K' is simple. Moreover, by Lemma 1.14 (ibid.), $\vdash_{\mathsf{K}} = \vdash_{\mathsf{K}'}$. Finally, we show that, for each $\mathcal{A} \in \mathsf{K}$, $\mathcal{A}/\theta$, where $\theta \triangleq \mathfrak{S}(\mathcal{A})$, is both λ-prime and -implicative, as well as $\mathcal{A}$ is. For consider arbitrary $p(\bar{a}), r(\bar{b}) \in \mathrm{At}_L(\mathfrak{A}/\theta)$. As $\nu_\theta : A \to (A/\theta)$ is surjective, there are some $\bar{a'} \in A^{\mu_L(p)}$ and $\bar{b'} \in A^{\mu_L(r)}$ such that $\nu_\theta \bar{a'} = \bar{a}$ and $\nu_\theta \bar{b'} = \bar{b}$. Recall that $\nu_\theta \in \mathrm{SHom}(\mathcal{A}, \mathcal{A}/\theta)$. In particular, $\mathcal{A} \models p(\bar{a'}) \Leftrightarrow (\mathcal{A}/\theta) \models p(\bar{a})$ and $\mathcal{A} \models r(\bar{b'}) \Leftrightarrow (\mathcal{A}/\theta) \models r(\bar{b})$. Moreover, $\mathcal{A} \models \lambda(p, r)[h] \Leftrightarrow (\mathcal{A}/\theta) \models \lambda(p, r)[\nu_\theta \circ h]$, where $h : V^{\mu_L(p) + \mu_L(r)} \cup V^1 \to A$, defined by:

$$
h x_i \triangleq \begin{cases} a'_i & \text{if } i \in \mu_L(p), \\ b'_{i - \mu_L(p)} & \text{if } i \in ((\mu_L(p) + \mu_L(r)) \setminus \mu_L(p)), \\ c & \text{elsewhere,} \end{cases}
$$

for all $i \in ((\mu_L(p) + \mu_L(r)) \cup 1$, where $c \in A$. This completes the argument of (v). $\qquad\qquad \square$

Corollary 2.5. *Let $\mathbb{T}$ be a finitary universal Horn L-theory, and both λ and λ' binary L-systems. Suppose $\mathbb{T}$ has DT with respect to both λ and λ'. Then, $\mathbb{T}$ holds PL with respect to λ iff it holds PL with respect to λ'.*

Proof. In that case, by Proposition 2.1, we have $\lambda \cong_{\mathbb{T}} \lambda'$. Hence, as both λ and λ' are purely-relational, any $\mathcal{A} \in \mathrm{Mod}(\mathbb{T})$ is compatible with λ iff it is compatible with λ', while $\lambda^{-1}[\mathcal{A}] = (\lambda')^{-1}[\mathcal{A}]$, in which case $\mathcal{A}$ is λ(-implicative)[-prime] iff it is λ'(-implicative)[-prime]. Then, Theorem 2.4(i)$\Leftrightarrow$(ii) completes the argument. $\qquad\square$

In this way, Peirce Law is an absolute property, holding regardless to any concrete Deduction Theorem schema.

Let λ be a finitary binary L-system. Given any $r, r' \in R$, choose any $\overline{\lambda(r, r')} \in \lambda(r, r')^n$, where $n \triangleq |\lambda(r, r')|$, such that $\mathrm{img}\, \overline{\lambda(r, r')} = \lambda(r, r')$. Then, given any $\overline{\varphi} \in \mathrm{Tm}_F(V^\infty)^{\mu_L(r)}$ and $\overline{\psi} \in \mathrm{Tm}_F(V^\infty)^{\mu_L(r')}$, put:

$$\overline{\lambda(r(\overline{\varphi}), r'(\overline{\psi}))} \triangleq \overline{\lambda(r, r')}[x_i/\varphi_i, x_{j+\mu_L(r)}/\psi_j]_{i\in\mu_L(r), j\in\mu_L(r')}.$$

Define the binary L-system $\mho_\lambda$ by:

$$(2.6) \qquad\qquad \mho_\lambda(\Phi, \Psi) \triangleq \lambda(\overline{\lambda(\Phi, \Psi)}, \Psi),$$

for all $\Phi, \Psi \in \mathrm{At}_L(V^\infty)$. In the sentential case, when in addition λ is a singleton, (2.6) means the standard definition of disjunction:

$$\mho_\lambda(x_0, x_1) = \lambda(\lambda(x_0, x_1), x_1).$$

Remark 2.6. Let λ be as above. Then, any λ-implicative and -prime L-structure $\mathcal{A}$ is $\mho_\lambda$-*disjunctive* in the sense of Appendix C of [43], i.e., it is compatible with $\mho_\lambda$, while, for all $\Phi, \Psi \in \mathrm{At}_L(\mathfrak{A})$, $\{\Phi, \Psi\} \cap (\mathcal{A}\downarrow) \neq \varnothing \Leftrightarrow (\mho_\lambda)^{-1}[\mathcal{A}] \models \langle\Phi, \Psi\rangle$. $\qquad\square$

Theorem 2.4 and Remark 2.6 immediately yield:

Corollary 2.7. *Let $\mathbb{T}$ be a finitary universal Horn L-theory and λ a binary L-system. Suppose $\mathbb{T}$ has DT and holds PL with respect to λ (in which case λ can be taken to be finitary, in view of both Note B.1 of [43] and Corollary 2.5). Then, $\vdash_{\mathbb{T}}$ is the consequence of a [the] class of [all] (simple) $\mho_\lambda$-disjunctive models of $\mathbb{T}$.*

2.2   Translations of Deduction Theorem schemes

Let τ and ρ be as in Definition 2.1 of [42]. Assume ρ is finitary. Given a binary L_1-system λ, define the binary L_2-system $\tau \bullet \lambda \star \rho$ as follows. Consider any $r, r' \in R_2$. Take any $\overline{\rho(r)} \in \rho(r)^n$, where $n \triangleq |\rho(r)| \in \omega$, such that $\mathrm{img}\, \overline{\rho(r)} = \rho(r)$. Put $(\lambda \star \rho)(r, r') \triangleq \lambda(\overline{\rho(r)}, \rho(r')[x_j/x_{j+\mu_{L_2}(r)}]_{j \in \mu_{L_2}(r')})$ and $(\tau \bullet \lambda \star \rho)(r, r') \triangleq \tau[(\lambda \star \rho)(r, r')]$.

Theorem 2.8. *Let $\mathbb{T}_1$, $\mathbb{T}_2$, τ and ρ be as in Definition 2.1 of [42]. Assume $\mathbb{T}_1$ and $\mathbb{T}_2$ are finitary and equivalent with respect to τ and ρ (in which case ρ can be taken to be finitary, in view of Proposition 2.3(i) of [42]). Suppose $\mathbb{T}_1$ has DT[and holds PL] with respect to a binary L_1-system λ. Then, $\mathbb{T}_2$ has DT[and holds PL] with respect to $\tau \bullet \lambda \star \rho$.*

Proof. The part with DT alone (viz., without involving PL) is due to Theorem A.6 of [44]. Now assume $\mathbb{T}_1$ has DT and holds PL with respect to λ. Then, by Theorem 2.4(i)$\Rightarrow$(iii), $\mathbb{T}_1 = \vdash_{\mathsf{K}_1}$, where K_1 is the class of all simple λ-prime and λ-implicative models of $\mathbb{T}_1$. On the other hand, by Theorem 2.10(i)$\Rightarrow$(iv) of [42], $\mathrm{Mod}_{\mathfrak{S}}(\mathbb{T}_2)$ is compatible with τ, $\mathrm{Mod}_{\mathfrak{S}}(\mathbb{T}_1)$ is compatible with ρ (and so is K_1), while the mappings τ^{-1} with domain $\mathrm{Mod}_{\mathfrak{S}}(\mathbb{T}_2)$ and ρ^{-1} with domain $\mathrm{Mod}_{\mathfrak{S}}(\mathbb{T}_1)$ are inverse to one another bijections between the classes involved. Put $\mathsf{K}_2 \triangleq \rho^{-1}[\mathsf{K}_1]$. Then, $\mathsf{K}_2 \subseteq \mathrm{Mod}_{\mathfrak{S}}(\mathbb{T}_2)$, so it is compatible with τ. Then, by Theorem 2.10(v)$\Rightarrow$(i), $\mathbb{T}_1$ and $\vdash_{\mathsf{K}_2}$ are equivalent with respect to τ and ρ, so $\mathbb{T}_2 = \vdash_{\mathsf{K}_2}$. Moreover, by Note 2.3 and the already-proved part without PL, every member of K_2 is $(\tau \bullet \lambda \star \rho)$-implicative, and so is compatible with $(\tau \bullet \lambda \star \rho)$. Thus, in view of Theorem 2.4(iv)$\Rightarrow$(i), it only remains to argue that every member of K_2 is $(\tau \bullet \lambda \star \rho)$-prime. For consider any $\mathcal{A} \in \mathsf{K}_1$ and arbitrary $p(\overline{a}), r(\overline{b}) \in \mathrm{At}_L(\rho^{-1}[\mathfrak{A}])$. Assume $\rho^{-1}[\mathcal{A}] \not\models p(\overline{a})$. Then, by (2.11) of [42], $\mathcal{A} \not\models \rho(p)[h]$, where $h : V^{\mu_L(p)+\mu_L(r)} \cup V^1 \to A$, defined by:

$$
h x_i \triangleq \begin{cases} a_i & \text{if } i \in \mu_L(p), \\ b_{i-\mu_L(p)} & \text{if } i \in ((\mu_L(p) + \mu_L(r)) \setminus \mu_L(p)), \\ c & \text{elsewhere,} \end{cases}
$$

for all $i \in ((\mu_L(p) + \mu_L(r)) \cup 1$, where $c \in A$. Hence, there is some $\Phi \in \rho(p)$ such that $\mathcal{A} \not\models \Phi[h]$. Then, as $\mathcal{A}$ is λ-prime, we have $\mathcal{A} \models \lambda(\Phi, (\lambda \star \rho)(p, r))[h]$. Therefore, by Proposition 2.2, $\mathcal{A} \models (\lambda \star \rho)(p, r)[h]$. On the other hand, $\mathcal{A} = \tau^{-1}[\rho^{-1}[\mathcal{A}]]$. Hence, by (2.11) of [42], $\rho^{-1}[\mathcal{A}] \models (\tau \bullet \lambda \star \rho)(p, r)[h]$. Thus, $(\tau \bullet \lambda \star \rho)^{-1}[\rho^{-1}[\mathcal{A}]] \models \langle p(\overline{a}), r(\overline{b}) \rangle$, as required. $\square$

Here, we take the opportunity to specify the second half of the comment after the proof of Theorem A.6 of [44], namely:

> the construction of appropriate binary systems involves solely relational components of translations under consideration, like such was the case, when dealing with purely relational translations alone

To be more precise, $\tau \bullet \lambda \star \rho$ does not involve the functional component of ρ but does involve that of τ.

2.3 Algebraic counterparts of both Deduction Theorem and Peirce Law

Let F and F' be functional signatures.

Given any binary $(F + \approx)$-system λ (that is naturally identified with the set $\lambda(\approx, \approx) \subseteq \mathrm{Eq}_F(V^4)$), a [pre]variety K of F-algebras is said to *have restricted equationally definable principal [relative] congruences* {*REDP[R]C*, for short} *(with respect to λ)*, provided, for each $\mathfrak{A} \in \mathsf{K}$ and every $\bar{a} \in A^4$, it holds that $(\langle a_2, a_3 \rangle \in \mathrm{Cg}_\mathsf{K}^\mathfrak{A}(\langle a_0, a_1 \rangle)) \Leftrightarrow (\mathfrak{A} + \approx) \models \lambda[x_i/a_i]_{i \in 4}$. (In case K is a variety, this concept is due to [8].) In view of (1.2) and Proposition 1.5, it is equivalent to the fact $\vdash_\mathsf{K}^\approx$ has ADT with respect to λ over $\mathrm{Mod}_3(\vdash_\mathsf{K}^\approx)$. Hence, Theorem B.3(i)$\Leftrightarrow$(iii) of [43] immediately yields:

Proposition 2.9. *Let K and λ be as above. Then, K has REDPRC with respect to λ iff $\vdash_\mathsf{K}^\approx$ has DT with respect to λ.*

As an immediate consequence of Corollary 2.13(iii)$\Rightarrow$(ii) of [42], Theorem 2.8 and Proposition 2.9, we first have:

Corollary 2.10. *Let K and K' be quasivarieties of, resp., F- and F'-algebras rationally equivalent with respect to functional (F, F')-translation τ and (F', F)-translation ρ, while λ as above. Suppose K has REDPRC with respect to λ. Then, K' has REDPRC with respect to $\tau[\lambda]$.*

Recall that, according to Corollary 1.25 of [42], $\vdash_\mathsf{K}^\approx$ is axiomatized by $\mathbb{T} = \mathbb{E}_F \cup \mathbb{A}$, where $\mathbb{A}$ is any axiomatization of K. This together with Theorem B.3 of [43] and Proposition 2.9 mean that any purely algebraic proof of REDPRC requires checking the conditions B.3(iii)**a)-d)** of [43] for such $\mathbb{T}$. The latter task is normally too complicated and, anyway, require *ad hoc* argumentation.[1] On

---

[1]It is then not surprising that original proofs of such results (cf. [14], [17], [46]) proved so complicated and smooth.

the other hand, the proof Theorem B.7 of [43] actually checking the conditions B.3(iii)**a)-d)** therein for enlargable Gentzen-style multiple-conclusion propositional calculi with structural rules is uniform and quite easy. Likewise, checking the conditions involved for implication and Hilbert-style calculi constituted by intuitionistic rules for implication (that is, (MP), (K) and (D)) and additional axioms is practically trivial. These observations make the following immediate consequence of Theorem A.6 of [44] (viz., the PL-less case of Theorem 2.8) and Proposition 2.9 more than acute for Universal Algebra:

Corollary 2.11. *Let* $\mathbb{T}$ *be a finitary universal Horn L-theory and* K *a quasivariety of* F'*-algebras being an equivalent purely-algebraic semantics for* $\mathbb{T}$*. Then,* $\mathbb{T}$ *has DT iff* K *has REDPRC.*

According to [41], a binary $(F + \approx)$-system λ is called an *implication system for* a class of of F-algebras K, provided each member of $K + \approx$ is both λ-implicative and -prime, in which case every member of K is said to be λ-*implicational*, while any algebra embeddable in it is λ-implicational as well.

Lemma 2.12. *Let* λ *be as above and* K *a prevariety of* F-*algebras having REDPRC with respect to* λ*. Then, any member of* K *is* λ-*implicational iff it is algebraically* K-*simple.*

Proof. Take any $\mathfrak{A} \in$ K. In view of Proposition 2.9, Note 2.3 and (1.2), $\mathfrak{A} + \approx$ is λ-implicative. In this way, $\mathfrak{A}$ is λ-implicational iff $\mathfrak{A} + \approx$ is λ-prime. First, assume $\mathfrak{A}$ is algebraically K-simple. Consider any $\bar{a} \in A^4$. Suppose $a_0 \neq a_1$, in which case $\mathrm{Cg}_{\mathsf{K}}^{\mathfrak{A}}(\langle a_0, a_1 \rangle) \neq \Delta_A$, so $\mathrm{Cg}_{\mathsf{K}}^{\mathfrak{A}}(\langle a_0, a_1 \rangle) = A^2$. In particular, $\langle a_2, a_3 \rangle \in \mathrm{Cg}_{\mathsf{K}}^{\mathfrak{A}}(\langle a_0, a_1 \rangle)$. Hence, $(\mathfrak{A} + \approx) \models \lambda[x_i/a_i]_{i \in 4}$. Thus, $\mathfrak{A} + \approx$ is λ-prime. Conversely, suppose $\mathfrak{A} + \approx$ is λ-prime. Take any $\theta \in \mathrm{Con}_{\mathsf{K}}(\mathfrak{A})$. Assume $\theta \neq \Delta_A$. Then, there is some $\bar{a} \in A^2 \cap \theta$ such that $a_0 \neq a_1$. Consider arbitrary $a_2, a_3 \in A$. Then, $(\mathfrak{A} + \approx) \models \lambda[x_i/a_i]_{i \in 4}$. Therefore, $\langle a_2, a_3 \rangle \in \mathrm{Cg}_{\mathsf{K}}^{\mathfrak{A}}(\langle a_0, a_1 \rangle) \subseteq \theta$. Thus, $\theta = A^2$, so $\mathfrak{A}$ is algebraically K-simple, as required. $\square$

A prevariety is said to be λ-*implicative* (cf. [41]), provided it is generated by a class of λ-implicational algebras.

As an immediate consequence of Lemma 2.12, we have:

Theorem 2.13. *Let* λ *be as above and* K *a prevariety of* F-*algebras having REDPRC with respect to* λ*. Then,* K *is* λ-*implicative iff it is generated by its algebraically relatively simple members.*

Theorem 2.4(i)$\Leftrightarrow$(v), Remark 1.4 and (1.2) directly imply:

23

Proposition 2.14. *Let* λ *be as above and* K *a quasivariety of* F*-algebras. Then,* K *is* λ*-implicative iff* $\vdash_{\mathsf{K}}^{\approx}$ *has DT and holds PL with respect to* λ.

By Propositions 2.9, 2.14 and Theorem 2.13, we immediately get:

Corollary 2.15. *Let* λ *be as above and* K *a quasivariety of* F*-algebras. Then,* K *is* λ*-implicative iff the following conditions hold:*

(i) K *has REDPRC with respect to* λ;

(ii) K *is generated by its algebraically relatively simple members.*

As an immediate consequence of Corollary 2.13(iii)$\Rightarrow$(ii) of [42], Theorem 2.8 and Proposition 2.14, we also have:

Corollary 2.16. *Let* K *and* K$'$ *be quasivarieties of, resp.,* F*- and* F'*-algebras rationally equivalent with respect to functional* (F, F')*-translation* τ *and* (F', F)*-translation* ρ*, while* λ *as above. Suppose* K *is* λ*-implicative. Then,* K$'$ *is* $\tau[\lambda]$*-implicative.*

After all, combining Theorem 2.8 and Proposition 2.14, we eventually obtain:

Corollary 2.17. *Let* $\mathbb{T}$ *be a finitary universal Horn* L*-theory and* K *a quasivariety of* F'*-algebras being an equivalent purely-algebraic semantics for* $\mathbb{T}$*. Then,* $\mathbb{T}$ *has DT and holds PL iff* K *is implicative.*

2.4 Deduction Theorem and Peirce Law for enlargable multiple-conclusion sequent calculi with structural rules

Throughout this and the next sections, fix an arbitrary first-order language L and any $\alpha, \beta \in \{\omega, \omega \setminus 1\}$. Put $\mathrm{Seq}_L^{[\alpha,\beta]}(O) \triangleq \mathrm{At}_{L[\alpha,\beta]}(O)$, where O is either V^{α}, for some $\alpha \leqslant \infty$, or an F-algebra.

Structural rules are the following $\langle \varnothing, R \rangle$-sequent rules of type $[\omega, \omega]$:

$$\text{(Reflexivity)} \qquad \Phi \rightarrowtail \Phi$$

$$\text{(Cut)} \qquad \frac{\Lambda, \Gamma \rightarrowtail \Delta, \Phi \quad \Phi, \Gamma \rightarrowtail \Delta, \Theta}{\Lambda, \Gamma \rightarrowtail \Delta, \Theta}$$

	Left	Right
(Enlargement)	$\dfrac{\Gamma \rightarrowtail \Delta}{\Phi, \Gamma \rightarrowtail \Delta}$	$\dfrac{\Gamma \rightarrowtail \Delta}{\Gamma \rightarrowtail \Delta, \Phi}$
(Permutation)	$\dfrac{\Lambda, \Phi, \Psi, \Gamma \rightarrowtail \Delta}{\Lambda, \Psi, \Phi, \Gamma \rightarrowtail \Delta}$	$\dfrac{\Gamma \rightarrowtail \Delta, \Phi, \Psi, \Theta}{\Gamma \rightarrowtail \Delta, \Psi, \Phi, \Theta}$

where $\Phi, \Psi \in \mathrm{At}_{\langle \varnothing, R \rangle}(V^\omega)$, while $\Gamma, \Lambda, \Delta, \Theta \in \mathrm{At}_{\langle \varnothing, R \rangle}(V^\omega)^*$.

Remark 2.18. The following (secondary) structural rules are derivable from the above (primary) structural rules:

$$\text{(Sharing)} \qquad \Lambda, \Phi, \Gamma \rightarrowtail \Delta, \Phi, \Theta$$

$$\begin{array}{ccc} & \text{Left} & \text{Right} \\ \text{(Contraction)} & \dfrac{\Phi, \Phi, \Gamma \rightarrowtail \Delta}{\Phi, \Gamma \rightarrowtail \Delta} & \dfrac{\Gamma \rightarrowtail \Delta, \Phi, \Phi}{\Gamma \rightarrowtail \Delta, \Phi} \end{array}$$

First, the derivability of Sharing from Reflexivity is by Enlargement and Permutation. Next, the derivability of Left Contraction is argued by the following demonstration (cf. Note 4.1 of [28] for the sentential case):

1. $\Phi, \Phi, \Gamma \rightarrowtail \Delta$, Hypothesis.

2. $\Phi, \Gamma \rightarrowtail \Delta, \Phi$, Sharing.

3. $\Phi, \Gamma \rightarrowtail \Delta$, Cut: 1,2.

Likewise, the derivability of Right Contraction is shown by symmetry. $\qquad \square$

The $\langle \varnothing, R \rangle$-sequent calculus of type $[\alpha, \beta]$ constituted by all structural rules of the same type is denoted by $\mathcal{S}_R^{[\alpha, \beta]}$.

Given any $(\overline{\Phi} \rightarrowtail \overline{\Psi}), (\overline{\Omega} \rightarrowtail \overline{\Upsilon}) \in \mathrm{Seq}_L^{[\alpha, \beta]}(O)$, where O is as above, put:

$$\begin{aligned}
((\overline{\Phi} \rightarrowtail \overline{\Psi}) \sqsubseteq (\overline{\Omega} \rightarrowtail \overline{\Upsilon})) & \Leftrightarrow & (((\mathrm{img}\,\overline{\Phi}) \subseteq (\mathrm{img}\,\overline{\Omega})) \\
 & \& & ((\mathrm{img}\,\overline{\Psi}) \subseteq (\mathrm{img}\,\overline{\Upsilon}))), \\
((\overline{\Phi} \rightarrowtail \overline{\Psi}) \simeq (\overline{\Omega} \rightarrowtail \overline{\Upsilon})) & \Leftrightarrow & (((\overline{\Omega} \rightarrowtail \overline{\Upsilon}) \sqsubseteq (\overline{\Phi} \rightarrowtail \overline{\Psi})) \\
 & \& & ((\overline{\Phi} \rightarrowtail \overline{\Psi}) \sqsubseteq (\overline{\Omega} \rightarrowtail \overline{\Upsilon}))), \\
((\overline{\Phi} \rightarrowtail \overline{\Psi}) \sqcup (\overline{\Omega} \rightarrowtail \overline{\Upsilon})) & \triangleq & ((\overline{\Phi}, \overline{\Omega}) \rightarrowtail (\overline{\Psi}, \overline{\Upsilon})).
\end{aligned}$$

Remark that:

(2.7) $$(\Phi \sqsubseteq \Psi) \Rightarrow (\Phi \vdash_{\mathcal{S}_R^{[\alpha, \beta]}} \Psi),$$

for all $\Phi, \Psi \in \mathrm{Seq}_L^{[\alpha, \beta]}(V^\infty)$.

An L-sequent calculus $\mathbb{G}$ of type $[\alpha, \beta]$ is said to be *enlargable*, provided, for every $(\Gamma \rightarrow \Phi) \in \mathbb{G}$ and each $\phi \in \mathrm{At}_L(V^\infty)$, both $\{(\phi, \Psi) | \Psi \in \Gamma\} \rightarrow (\phi, \Phi)$ and $\{(\Psi, \phi) | \Psi \in \Gamma\} \rightarrow (\Phi, \phi)$ are derivable in $\mathbb{G}$. In the sentential case, such calculi were referred to as *disjunctive* in [32].

By $\lambda_{L^{[\alpha,\beta]}}$ we denote the binary $L^{[\alpha,\beta]}$-system, given by (cf. Appendix B of [43]):

$$(2.8) \qquad \lambda_{L^{[\alpha,\beta]}}(\overline{\varphi} \rightarrowtail \overline{\psi}, \overline{\xi} \rightarrowtail \overline{\eta}) = \{\psi_i, \overline{\xi} \rightarrowtail \overline{\eta} | i \in l\}$$
$$\cup \; \{\overline{\xi} \rightarrowtail \overline{\eta}, \varphi_j | j \in k\},$$

where $\overline{\xi} \in \mathrm{At}_L(V^\infty)^m, \overline{\eta} \in \mathrm{At}_L(V^\infty)^n, \overline{\varphi} \in \mathrm{At}_L(V^\infty)^k, \overline{\psi} \in \mathrm{At}_L(V^\infty)^l$, while $m, k \in \alpha$, whereas $n, l \in \beta$.

Theorem 2.19. *Let $\mathbb{G}$ be as above. Suppose it is enlargable and contains all structural rules of type $[\alpha, \beta]$, Then, it has DT and holds PL with respect to $\lambda_{L^{[\alpha,\beta]}}$.*

Proof. The part with DT alone (without PL) is due to Theorem B.7 of [43] (cf. Theorem 4.2 of [28] for the sentential case). For proving (2.1), consider any $\Phi, \Psi \in \mathrm{Seq}_L^{[\alpha,\beta]}(V^\infty)$. Take any $\Omega \in \lambda_{L^{[\alpha,\beta]}}(\Phi, \Psi)$ (cf. (2.8)). It is easy to see that

$$\lambda_{L^{[\alpha,\beta]}}(\Omega, \Phi) \ni \Upsilon \simeq \Phi.$$

Then, (2.7) completes the argument. $\qquad \square$

Proposition 2.20. $\mho_{\lambda_{L^{[\alpha,\beta]}}}(\Phi, \Psi) \dashv\vdash_{S_R^{[\alpha,\beta]}} (\Phi \sqcup \Psi)$, *for all* $\Phi, \Psi \in \mathrm{Seq}_L^{[\alpha,\beta]}$.

Proof. Using (2.6) and (2.8), it is routine checking that any element of

$$\mho_{\lambda_{L^{[\alpha,\beta]}}}(\Phi, \Psi)$$

is either a Sharing substitutional instance or $\simeq (\Phi \sqcup \Psi)$. Then, (2.7) and Remark 2.18 complete the argument. $\qquad \square$

An $L^{[\alpha,\beta]}$-structure $\mathcal{A}$ is said to be $\sqcup$-*disjunctive*, provided:

$$\{\Phi, \Psi\} \cap (\mathcal{A} \downarrow) \neq \varnothing \Leftrightarrow \mathcal{A} \models (\Phi \sqcup \Psi),$$

for all $\Phi, \Psi \in \mathrm{Seq}_L(\mathfrak{A})$.

By Proposition 2.20, we immediately get:

Corollary 2.21. *Any model of $S_R^{[\alpha,\beta]}$ is $\mho_{\lambda_{L^{[\alpha,\beta]}}}$-disjunctive iff it is $\sqcup$-disjunctive.*

2.5 Semantics of enlargable multiple-conclusion sequent calculi

Given an L-structure $\mathcal{A}$, define the $L^{[\alpha,\beta]}$-structure $\mathcal{A}^{[\alpha,\beta]}$ with the same underlying algebra as follows: for all $\Phi = (\overline{\phi} \rightarrowtail \overline{\psi}) \in \mathrm{Seq}_L(\mathfrak{A})$, put $\mathcal{A}^{[\alpha,\beta]} \models \Phi \Leftrightarrow (((\mathrm{img}\,\overline{\phi}) \subseteq (\mathcal{A} \downarrow)) \Rightarrow ((\mathrm{img}\,\overline{\phi}) \cap (\mathcal{A} \downarrow) \neq \varnothing))$. Given a class of L-structures K, set $\mathsf{K}^{[\alpha,\beta]} \triangleq \{\mathcal{A}^{[\alpha,\beta]} | \mathcal{A} \in \mathsf{K}\}$.

Remark 2.22. It is straightforward to verify that:

$$\mathrm{SHom}(\mathcal{A}, \mathcal{B}) \subseteq \mathrm{SHom}(\mathcal{A}^{[\alpha,\beta]}, \mathcal{B}^{[\alpha,\beta]}),$$

for any two L-structures $\mathcal{A}$ and $\mathcal{B}$. $\qquad\square$

Theorem 2.23. *Let $\mathbb{G}$ be an L-sequent calculus of type $[\alpha, \beta]$. Then, the following are equivalent:*

(i) $\mathbb{G}$ *is enlargable, while all structural rules of the same type are derivable in $\mathbb{G}$;*

(ii) $\vdash_{\mathbb{G}} = \vdash_{\mathsf{K}^{[\alpha,\beta]}}$, *for some class K of L-structures.*

Proof. (ii)$\Rightarrow$(i) is straghtforward. For proving the converse, assume (i) holds. Consider the following two complementary cases:

1. $\alpha = \beta = \omega$.
   Then, by Theorem 2.19, Corollaries 2.7, 2.21 and (1.1), $\vdash_{\mathbb{G}} = \vdash_{\mathsf{M}}$, for some class M of proper $\sqcup$-disjunctive $L^{[\alpha,\beta]}$-structures. Consider any $\mathcal{A} \in \mathsf{M}$. Define the L-structure $\mathcal{A} \downarrow$ with underlying algebra $\mathfrak{A}$ as follows: for all $\phi \in \mathrm{At}_L(\mathfrak{A})$, put $(\mathcal{A} \downarrow) \models \phi \Leftrightarrow \mathcal{A} \models (\rightarrowtail \phi)$. First of all, we prove that:

   (2.9) $$\mathcal{A} \models (\phi \rightarrowtail) \Leftrightarrow \mathcal{A} \not\models (\rightarrowtail \phi),$$

   for all $\phi \in \mathrm{At}_L(\mathfrak{A})$. The fact that either $\mathcal{A} \models (\phi \rightarrowtail)$ or $\mathcal{A} \models (\rightarrowtail \phi)$ is by Reflexivity and the $\sqcup$-disjunctivity of $\mathcal{A}$. The fact that either $\mathcal{A} \not\models (\phi \rightarrowtail)$ or $\mathcal{A} \not\models (\rightarrowtail \phi)$ is by Cut, Enlargement and the non-totality of $\mathcal{A}$. Thus, (2.9) does hold. Finally, by (2.9) and the $\sqcup$-disjunctivity of $\mathcal{A}$, we conclude that $\mathcal{A} = (\mathcal{A} \downarrow)^{[\alpha,\beta]}$. In this way, setting $\mathsf{K} \triangleq \{(\mathcal{A} \downarrow) | \mathcal{A} \in \mathsf{M}\}$, we eventually get $\mathsf{M} = \mathsf{K}^{[\alpha,\beta]}$, as actually required.

2. $\alpha \cap \beta = \omega \setminus 1$.

Consider the L-sequent calculus $\mathbb{G}' \triangleq \mathbb{G} \cup \mathcal{S}_R^{[\omega,\omega]}$ of type $[\omega,\omega]$. As both $\mathbb{G}$ and $\mathcal{S}_R^{[\omega,\omega]}$ are enlargable, $\mathbb{G}'$ is enlargable as well. Hence, by the case 1, there is some class K of L-structures such that $\vdash_{\mathbb{G}'} = \vdash_{\mathsf{K}^{[\omega,\omega]}}$. Consider any L-sequent rule $\mathfrak{R} = (\Gamma \to \Phi)$ of type $[\alpha,\beta]$. Clearly, it is true in $\mathsf{K}^{[\omega,\omega]}$ iff it is true in $\mathsf{K}^{[\alpha,\beta]}$. Moreover, for any $(\Delta \to \Psi) \in \mathcal{S}_R^{[\omega,\omega]}$, $\Psi \in \mathrm{Seq}_L^{[\alpha,\beta]}(V^\omega)$, whenever $\Delta \subseteq \mathrm{Seq}_L^{[\alpha,\beta]}(V^\omega)$, in which case $(\Delta \to \Psi) \in \mathcal{S}_R^{[\alpha,\beta]}$. Hence, any $\mathbb{G}'$-derivation of Φ from Γ is a $(\mathbb{G} \cup \mathcal{S}_R^{[\alpha,\beta]})$-derivation of Φ from Γ. Therefore, $\mathfrak{R}$ is derivable in $\mathbb{G}$, whenever it is derivable in $\mathbb{G}'$. As the converse is obviously true, for $\mathbb{G} \subseteq \mathbb{G}'$, we thus conclude that $\vdash_{\mathbb{G}} = \vdash_{\mathsf{K}^{[\alpha,\beta]}}$, as required.

This completes the argument. $\qquad\qquad\qquad\qquad\qquad\qquad\qquad\square$

This extends Theorem 3.5 of [32] being the particular sentential case of Theorem 2.23.

2.6 Contraposable multiple-conclusion propositional sequent calculi

Here, F is supposed to be a sentential language (viz., a functional signature) containing a unary connective $\sim$ (*weak*, viz., *constructive negation*), while $\beta = \alpha$.

Let $\diamond$ be a binary connective in F. By induction on $n \in \omega \setminus 1$, for any $\overline{\phi} \in \mathrm{Tm}_F(V^\infty)^n$, define $\diamond(\overline{\phi})$ as follows:

$$\diamond(\overline{\phi}) \triangleq \phi_0 \quad (n = 1),$$
$$\diamond(\overline{\phi}) \triangleq ((\diamond(\overline{\phi} \restriction (n-1))) \diamond \phi_{n-1}) \quad (n \in \omega \setminus 2).$$

Let $\wr$ be a unary connective in F. Given any $\overline{\phi} \in \mathrm{Tm}_F(V^\infty)^n$, where $n \in \omega$, put $\wr\overline{\phi} \triangleq \langle \wr\phi_i \rangle_{i \in n} \in \mathrm{Tm}_F(V^\infty)^n$.

Then, given any $\Phi = (\overline{\varphi} \rightarrowtail \overline{\psi}) \in \mathrm{Seq}_F^{[\alpha,\alpha]}(V^\infty)$, put $\widetilde{\Phi} \triangleq (\sim\overline{\psi} \rightarrowtail \sim\overline{\varphi}) \in \mathrm{Seq}_F^{[\alpha,\alpha]}(V^\infty)$. Finally, given any $\Gamma \subseteq \mathrm{Seq}_F^{[\alpha,\alpha]}(V^\infty)$, put $\widetilde{\Gamma} \triangleq \{\widetilde{\Phi} | \Phi \in \Gamma\} \subseteq \mathrm{Seq}_F^{[\alpha,\alpha]}(V^\infty)$.

Given an F-sequent calculus $\mathbb{G}$ of type $[\alpha,\alpha]$, by $\widehat{\mathbb{G}}$ we denote the one resulted from the former by adding the single *Weak Contraposition* rule:

(2.10)
$$\frac{x_0 \rightarrowtail x_1}{\sim x_1 \rightarrowtail \sim x_0}.$$

Then, $\mathbb{G}$ is said to be *weakly contraposable*, provided, for each $\Gamma \rightarrowtail \Phi \in \mathbb{G}$, $\widetilde{\Gamma} \rightarrowtail \widetilde{\Phi}$ is derivable in $\mathbb{G}$, while the following rule is derivable in $\mathbb{G}$:

(2.11)
$$\frac{x_0 \rightarrowtail x_1 \quad \sim x_1 \rightarrowtail \sim x_0}{\sim\sim x_0 \rightarrowtail \sim\sim x_1},$$

In that case, $\mathbb{G}$ is said to be *strongly contraposable*, provided the following *Strong Contraposition* rule is derivable in $\widehat{\mathbb{G}}$:

(2.12)
$$\frac{\overline{x}(0, m) \rightarrowtail \overline{x}(m, n)}{\sim\overline{x}(m, n) \rightarrowtail \sim\overline{x}(0, m)},$$

where $m, n \in \alpha$.

Lemma 2.24 (Key Lemma). *Let $\mathbb{G}$ be as above. Assume it is strongly contraposable. Then, for any $(\Gamma \rightarrowtail \Phi) \in \vdash_{\widehat{\mathbb{G}}}$, it holds that $\Gamma \cup \widetilde{\Gamma} \vdash_{\mathbb{G}} \{\Phi, \widetilde{\Phi}\}$.*

Proof. By induction on the length of a $\widehat{\mathbb{G}}$-derivation of Φ from Γ. The case $\Phi \in \Gamma$ is trivial. Next, suppose there are some $\mathfrak{R} = (\Delta \rightarrow \Psi) \in \widehat{\mathbb{G}}$ and some F-substitution σ such that $\sigma\Psi = \Phi$, while $\Gamma \vdash_{\widehat{\mathbb{G}}} \sigma[\Delta]$, in which case, by induction hypothesis, we have $\Gamma \cup \widetilde{\Gamma} \vdash_{\mathbb{G}} \sigma[\Delta] \cup \widetilde{\sigma[\Delta]}$. Consider the following two complementary cases:

1. $\mathfrak{R} \in \mathbb{G}$.
   Then, both $\sigma[\Delta] \vdash_{\mathbb{G}} \Phi$ and $\widetilde{\sigma[\Delta]} = \sigma[\widetilde{\Delta}] \vdash_{\mathbb{G}} \sigma\widetilde{\Psi} = \widetilde{\Phi}$, so $\Gamma \cup \widetilde{\Gamma} \vdash_{\mathbb{G}} \{\Phi, \widetilde{\Phi}\}$, as required.

2. $\mathfrak{R} \notin \mathbb{G}$.
   Then $\mathfrak{R}$ is equal to (2.10), in which case $\sigma[\Delta] = \{\phi_0 \rightarrowtail \phi_1\}$, while $\Phi = (\sim\phi_1 \rightarrowtail \sim\phi_0)$, whereas $\phi_i = \sigma x_i$, for each $i \in 2$. Then, $\Phi \in \widetilde{\sigma[\Delta]}$, while $\sigma[\Delta] \cup \widetilde{\sigma[\Delta]} \vdash_{\mathbb{G}} \widetilde{\Phi}$, in view of (2.11). Hence, $\Gamma \cup \widetilde{\Gamma} \vdash_{\mathbb{G}} \{\Phi, \widetilde{\Phi}\}$ as well.

This completes the argument. $\qquad\square$

Let λ be a binary $F^{[\alpha,\alpha]}$-system. Define the one $\widetilde{\lambda}$ as follows: for all $k, l, m, n \in \alpha$, put:

$$\widetilde{\lambda}(\rightarrowtail_l^k, \rightarrowtail_n^m) \triangleq \lambda(\langle\overline{x}(0, k) \rightarrowtail \overline{x}(k, l), \sim\overline{x}(k, l) \rightarrowtail \sim\overline{x}(0, k)\rangle,$$
$$\{\overline{x}(k + l - 1, m) \rightarrowtail \overline{x}(k + l + m - 1, n);$$
$$\sim\overline{x}(k + l + m - 1, n) \rightarrowtail \sim\overline{x}(k + l - 1, m)\}).$$

To provide this (rather cumbersome) purely formal definition with more intuition, just notice that:

$$(2.13) \qquad \tilde{\lambda}(\Phi, \Psi) = \lambda(\langle \Phi, \tilde{\Phi} \rangle, \{\Psi, \tilde{\Psi}\}),$$

for all $\Phi, \Psi \in \mathrm{Seq}_F^{[\alpha,\alpha]}(V^\infty)$.

Theorem 2.25. *Let $\mathbb{G}$ and λ be as above. Suppose $\mathbb{G}$ is strongly cotraposable and has DT with respect to λ. Then, $\widehat{\mathbb{G}}$ had DT with respect to $\tilde{\lambda}$.*

Proof. Take any $\Gamma \cup \{\Phi, \Psi\} \subseteq \mathrm{Seq}_F^{[\alpha,\alpha]}(V^\infty)$, where Γ is a set. First, assume $\Gamma \cup \{\Phi\} \vdash_{\widehat{\mathbb{G}}} \Psi$. Then, by Lemma 2.24, we have $\Gamma \cup \tilde{\Gamma} \cup \{\Phi, \tilde{\Phi}\} \vdash_{\mathbb{G}} \{\Psi, \tilde{\Psi}\}$. Therefore, by (B.7) of [43] and (2.13), we get $\Gamma \cup \tilde{\Gamma} \vdash_{\mathbb{G}} \tilde{\lambda}(\Phi, \Psi)$, in which case $\Gamma \cup \tilde{\Gamma} \vdash_{\widehat{\mathbb{G}}} \tilde{\lambda}(\Phi, \Psi)$, for $\mathbb{G} \subseteq \widehat{\mathbb{G}}$. Moreover, by (2.12), $\Gamma \vdash_{\widehat{\mathbb{G}}} \Gamma \cup \tilde{\Gamma}$. Hence, $\Gamma \vdash_{\widehat{\mathbb{G}}} \tilde{\lambda}(\Phi, \Psi)$. Conversely, assume $\Gamma \vdash_{\widehat{\mathbb{G}}} \tilde{\lambda}(\Phi, \Psi)$. Then, by Lemma 2.24, we have $\Gamma \cup \tilde{\Gamma} \vdash_{\mathbb{G}} \tilde{\lambda}(\Phi, \Psi)$. Therefore, by (B.7) of [43] and (2.13), we get $\Gamma \cup \tilde{\Gamma} \cup \{\Phi, \tilde{\Phi}\} \vdash_{\mathbb{G}} \Psi$, in which case $\Gamma \cup \tilde{\Gamma} \cup \{\Phi, \tilde{\Phi}\} \vdash_{\widehat{\mathbb{G}}} \Psi$, for $\mathbb{G} \subseteq \widehat{\mathbb{G}}$. Moreover, by (2.12), $\Gamma \cup \{\Phi\} \vdash_{\widehat{\mathbb{G}}} \Gamma \cup \tilde{\Gamma} \cup \{\Phi, \tilde{\Phi}\}$. Hence, $\Gamma \cup \{\Phi\} \vdash_{\widehat{\mathbb{G}}} \Psi$. Thus, (B.2) of [43] does hold, as required. $\qquad \square$

As it is demonstrated in Section 3.3, the "Peirce Law strengthening" of Theorem 2.25 is not, generally speaking, valid. More precisely, Section 3.3 provides a counterexample of some strongly contraposable $\mathbb{G}$ that has DT and holds PL with respect to some λ, while $\widehat{\mathbb{G}}$ does not hold PL with respect to $\tilde{\lambda}$.

Combining (2.8) with (2.13), we get:

$$
\begin{aligned}
(2.14) \quad \widetilde{\lambda_{F[\alpha,\alpha]}}(\overline{\varphi} \rightarrowtail \overline{\psi}, \overline{\xi} \rightarrowtail \overline{\eta}) \;=\; & \{\sim\varphi_j, \psi_i, \overline{\xi} \rightarrowtail \overline{\eta} \,|\, i \in l, j \in k\} \\
\cup\; & \{\sim\varphi_j, \overline{\xi} \rightarrowtail \overline{\eta}, \varphi_\jmath \,|\, j, \jmath \in k\} \\
\cup\; & \{\psi_i, \overline{\xi} \rightarrowtail \overline{\eta}, \sim\psi_\imath \,|\, i, \imath \in l\} \\
\cup\; & \{\overline{\xi} \rightarrowtail \overline{\eta}, \varphi_j, \sim\psi_i \,|\, j \in k, i \in l\} \\
\cup\; & \{\sim\varphi_j, \psi_i, \sim\overline{\eta} \rightarrowtail \sim\overline{\xi} \,|\, i \in l, j \in k\} \\
\cup\; & \{\sim\varphi_j, \sim\overline{\eta} \rightarrowtail \sim\overline{\xi}, \varphi_\jmath \,|\, j, \jmath \in k\} \\
\cup\; & \{\psi_i, \sim\overline{\eta} \rightarrowtail \sim\overline{\xi}, \sim\psi_\imath \,|\, i, \imath \in l\} \\
\cup\; & \{\sim\overline{\eta} \rightarrowtail \sim\overline{\xi}, \varphi_j, \sim\psi_i \,|\, j \in k, i \in l\},
\end{aligned}
$$

where $\overline{\xi} \in \mathrm{Tm}_F(V^\infty)^m, \overline{\eta} \in \mathrm{Tm}_F(V^\infty)^n, \overline{\varphi} \in \mathrm{Tm}_F(V^\infty)^k, \overline{\psi} \in \mathrm{Tm}_F(V^\infty)^l$, while $m, n, k, l \in \alpha$.

By Theorems 2.19 and 2.25, we immediately obtain:

30

Corollary 2.26. *Let* $\mathbb{G}$ *be as above. Suppose* $\mathbb{G}$ *is strongly cotraposable, enlargable and contains all structural rules of type* $[\alpha, \alpha]$. *Then,* $\widehat{\mathbb{G}}$ *has DT with respect to* $\widetilde{\lambda_{F[\alpha,\alpha]}}$.

Like Theorem 2.25, the Peirce Law strengthening of Corollary 2.26 is not, generally speaking, valid, as it is equally demonstrated in Section 3.3. There is still a rather extensive kind of strongly contraposable, enlargable calculi, extensions of which by Weak Contraposition do hold Pierce Law. More precisely, we have:

Theorem 2.27. *Let* $\mathbb{G}$ *be as above. Suppose* $\mathbb{G}$ *is strongly cotraposable, enlargable and contains all structural rules of type* $[\alpha, \alpha]$, *while the rules inverse to the Strong Contrsposition ones* (2.12) *of the same type are derivable in* $\widehat{\mathbb{G}}$. *Then,* $\widehat{\mathbb{G}}$ *has DT and holds PL with respect to* $\widetilde{\lambda_{F[\alpha,\alpha]}}$.

Proof. The part of DT alone (viz., without PL) is due to Corollary 2.26. For proving (2.1), consider any $\Phi, \Psi \in \mathrm{Seq}_F^{[\alpha,\alpha]}(V^\infty)$. Take any $\Omega \in \widetilde{\lambda_{F[\alpha,\alpha]}}(\Phi, \Psi)$ (cf. (2.14)). It is easy to see that (cf. (2.8)):

$$\widetilde{\lambda_{F[\alpha,\alpha]}}(\langle \Omega, \widetilde{\Omega}\rangle, \widetilde{\Phi}) \ni \Upsilon \simeq \widetilde{\Phi}.$$

Then, (2.13), (2.7) and the rules inverse to the Strong Contrsposition ones (2.12) complete the argument. $\qquad\qquad\square$

Section 3.3 provides a counterexample showing that the condition of the reversibility of Strong Contraposition rules is essential for Theorem 2.27 to hold.

By *involution* rules we mean the following F-sequent rules of type $[\omega, \omega]$ (where $m, n \in \omega$):

(2.15)
$$\frac{\overline{x}(0, m+1) \rightarrowtail \overline{x}(m+1, n)}{\sim\sim x_0, \overline{x}(1, m) \rightarrowtail \overline{x}(m+1, n)},$$

(2.16)
$$\frac{\overline{x}(m+1, n) \rightarrowtail \overline{x}(0, m+1)}{\overline{x}(m+1, n) \rightarrowtail \sim\sim x_0, \overline{x}(1, m)}.$$

These are due to [23].

Remark 2.28. Under presence of structural rules, involution ones are reversible. More precisely, the following axioms are immediately derivable from Reflexivity by means of (2.15) and (2.16) of type $[\alpha, \alpha]$, respectively:

(2.17)
$$\sim\sim x_0 \;\rightarrowtail\; x_0,$$

(2.18)
$$x_0 \;\rightarrowtail\; \sim\sim x_0.$$

On the other hand, the rules inverse to involution ones of the same type (as well as the latter themselves) are derivable from the axioms (2.17) and (2.18) by means of structural rules of the same type. Moreover, these axioms are equally derivable from Reflexivity by means of the rules inverse to involution ones of the same type. In particular, involution rules of the same type and those which are inverse to them are, so to say, inter-derivable modulo structural rules of the same type. □

Lemma 2.29. *Let* $\mathbb{G}$ *be as above. Suppose structural, involution and strong contraposition rules of type* $[\alpha, \alpha]$ *are derivable in* $\widehat{\mathbb{G}}$. *Then, the rules inverse to Strong Contrasposition ones of the same type are derivable in* $\widehat{\mathbb{G}}$.

Proof. Consider any $\Phi \in \mathrm{Seq}_F^{[\alpha,\alpha]}(V^\infty)$. Then, by (2.12), we first have $\widetilde{\Phi} \vdash_{\widehat{\mathbb{G}}} \widetilde{\widetilde{\Phi}}$. Moreover, by the rules inverse to involution ones (cf. Remark 2.28), we also have $\widetilde{\widetilde{\Phi}} \vdash_{\widehat{\mathbb{G}}} \Phi$. Thus, $\widetilde{\Phi} \vdash_{\widehat{\mathbb{G}}} \Phi$, as required. □

Theorem 2.27 and Lemma 2.29 immediately yield:

Corollary 2.30. *Let* $\mathbb{G}$ *be as above. Suppose* $\mathbb{G}$ *is strongly cotraposable, enlargable and contains all structural and involution rules of type* $[\alpha, \alpha]$. *Then,* $\widehat{\mathbb{G}}$ *has DT and holds PL with respect to* $\widetilde{\lambda_{F[\alpha,\alpha]}}$.

Section 3.3 provides a counterexample showing that the condition of $\mathbb{G}$ containing involution rules is essential for Corollary 2.30 to hold.

Concluding this section, F is supposed to contain also binary connectives $\wedge$ and $\vee$ (*conjunction* and *disjunction*, respectively). The following are *classical* rules for those connectives, belonging to Gentzen's LK [9] (where $n \in \omega$, $m \in \alpha$):

$$(\wedge \rightarrowtail) \quad \frac{\overline{x}(0, n+2) \rightarrowtail \overline{x}(n+2, m)}{x_0 \wedge x_1, \overline{x}(2, n) \rightarrowtail \overline{x}(n+2, m)}$$

$$(\rightarrowtail \wedge) \quad \frac{\overline{x}(n+2, m) \rightarrowtail x_0, \overline{x}(2, n) \quad \overline{x}(n+2, m) \rightarrowtail \overline{x}(1, n+1)}{\overline{x}(n+2, m) \rightarrowtail x_0 \wedge x_1, \overline{x}(2, n)}$$

$$(\rightarrowtail \vee) \quad \frac{\overline{x}(n+2, m) \rightarrowtail \overline{x}(0, n+2)}{\overline{x}(n+2, m) \rightarrowtail x_0 \vee x_1, \overline{x}(2, n)}$$

$$(\vee \rightarrowtail) \quad \frac{x_0, \overline{x}(2, n) \rightarrowtail \overline{x}(n+2, m) \quad \overline{x}(1, n+1) \rightarrowtail \overline{x}(n+2, m)}{x_0 \vee x_1, \overline{x}(2, n) \rightarrowtail \overline{x}(n+2, m)}$$

Next, the following are *De Morgan* rules for all connectives under consideration, going back to [23] (where $n \in \omega$, $m \in \alpha$):

$$(\sim\!\vee \rightarrowtail) \qquad \frac{\sim\!x_0, \sim\!x_1, \overline{x}(2, n) \rightarrowtail \overline{x}(n+2, m)}{\sim\!(x_0 \vee x_1), \overline{x}(2, n) \rightarrowtail \overline{x}(n+2, m)}$$

$$(\rightarrowtail \sim\!\vee) \qquad \frac{\overline{x}(n+2, m) \rightarrowtail \sim\!x_0, \overline{x}(2, n) \quad \overline{x}(n+2, m) \rightarrowtail \sim\!x_1, \overline{x}(2, n)}{\overline{x}(n+2, m) \rightarrowtail \sim\!(x_0 \vee x_1), \overline{x}(2, n)}$$

$$(\rightarrowtail \sim\!\wedge) \qquad \frac{\overline{x}(n+2, m) \rightarrowtail \sim\!x_0, \sim\!x_1, \overline{x}(2, n)}{\overline{x}(n+2, m) \rightarrowtail \sim\!(x_0 \wedge x_1), \overline{x}(2, n)}$$

$$(\sim\!\wedge \rightarrowtail) \qquad \frac{\sim\!x_0, \overline{x}(2, n) \rightarrowtail \overline{x}(n+2, m) \quad \sim\!x_1, \overline{x}(2, n) \rightarrowtail \overline{x}(n+2, m)}{\sim\!(x_0 \wedge x_1), \overline{x}(2, n) \rightarrowtail \overline{x}(n+2, m)}$$

Remark 2.31. Under presence of structural rules, classical and De Morgan rules are reversible. More precisely, the following axioms are derivable from Sharing by means of classical rules of type $[\alpha, \alpha]$:

$$(2.19) \qquad (x_0 \wedge x_1) \;\rightarrowtail\; x_0,$$
$$(2.20) \qquad (x_0 \wedge x_1) \;\rightarrowtail\; x_1,$$
$$(2.21) \qquad x_0, x_1 \;\rightarrowtail\; (x_0 \wedge x_1),$$
$$(2.22) \qquad x_0 \;\rightarrowtail\; (x_0 \vee x_1),$$
$$(2.23) \qquad x_1 \;\rightarrowtail\; (x_0 \vee x_1),$$
$$(2.24) \qquad (x_0 \vee x_1) \;\rightarrowtail\; x_0, x_1.$$

On the other hand, the rules inverse to classical ones of the same type (as well as the latter themselves) are derivable from the axioms (2.19)—(2.24) by means of structural rules of the same type. Moreover, these axioms are equally derivable from Reflexivity by means of the rules inverse to classical ones of the same type. In particular, classical rules of the same type and those which are inverse to them are, so to say, inter-derivable modulo structural rules of the same type. Likewise, the following axioms are derivable from Sharing by means of De

Morgan rules of the same type:

$$(2.25) \qquad \sim(x_0 \vee x_1) \rightarrowtail \sim x_0,$$

$$(2.26) \qquad \sim(x_0 \vee x_1) \rightarrowtail \sim x_1,$$

$$(2.27) \qquad \sim x_0, \sim x_1 \rightarrowtail \sim(x_0 \vee x_1),$$

$$(2.28) \qquad \sim x_0 \rightarrowtail \sim(x_0 \wedge x_1),$$

$$(2.29) \qquad \sim x_1 \rightarrowtail \sim(x_0 \wedge x_1),$$

$$(2.30) \qquad \sim(x_0 \wedge x_1) \rightarrowtail \sim x_0, \sim x_1.$$

On the other hand, the rules inverse to De Morgan ones of the same type (as well as the latter themselves) are derivable from the axioms (2.25)—(2.30) by means of structural rules of the same type. Moreover, these axioms are equally derivable from Reflexivity by means of the rules inverse to De Morgan ones of the same type. In particular, De Morgan rules of the same type and those which are inverse to them are, so to say, inter-derivable modulo structural rules of the same type. $\square$

From now on, it is supposed that F contains *truth* and *falsehood* constants $\perp$ and $\top$, whenever $\alpha = \omega$. Then, an F-sequent calculus $\mathbb{G}$ of type $[\alpha, \alpha]$ is referred to as *normal*, provided the following axioms are derivable in it, whenever $\alpha = \omega$:

$$(2.31) \qquad \sim\top \rightarrowtail,$$

$$(2.32) \qquad \rightarrowtail \sim\perp.$$

Lemma 2.32. *Let $\mathbb{G}$ be as above. Suppose it is weakly contraposable, normal and contains all structural, classical and De Morgan rules. Then, it is strongly contraposable.*

Proof. Take any $m, n \in \alpha$. Put:

$$\overline{\varphi} \triangleq \begin{cases} \top & \text{if } m = 0, \\ \overline{x}(0, m) & \text{otherwise.} \end{cases}$$

$$\overline{\psi} \triangleq \begin{cases} \perp & \text{if } n = 0, \\ \overline{x}(m, n) & \text{otherwise.} \end{cases}$$

By double induction on m and n and both classical and structural rules together with (2.10), the rule:

$$\frac{\overline{x}(0, m) \rightarrowtail \overline{x}(m, n)}{\sim \vee \overline{\psi} \rightarrowtail \sim \wedge \overline{\varphi}}$$

34

is derivable in $\widehat{\mathbb{G}}$. Then, applying double induction on m and n collectively with Remark 2.31 for De Morgan rules and both (2.31) and (2.32) together with structural rules, we conclude that (2.12) is derivable in $\widehat{\mathbb{G}}$, as required. $\qquad\square$

Lemma 2.32 collectively with Corollaries 2.26 and 2.30 immediately yield:

Corollary 2.33. *Let $\mathbb{G}$ be as above. Suppose $\mathbb{G}$ is weakly cotraposable, enlargable, normal and contains all [involution,]structural, classical and De Morgan rules of type $[\alpha, \alpha]$. Then, $\widehat{\mathbb{G}}$ has DT[and holds PL] with respect to $\widetilde{\lambda}_{F[\alpha,\alpha]}$.*

In Section 3.3 we provide a counterexample, showing that involution rules cannot be disregarded in the formulation of Corollary 2.33 for PL to hold.

Chapter 3

Examples

First of all, Corollary 2.11 [2.17] is applied to any finitely-algebraizable ω-compact sentential logics with DT[and PL]. This includes the [classical], intuitionistic as well as all intermediate (including *linear*; cf. [6]) logics, in which case we get REDPC in Heiting algebras [and the implicativity of the variety of Boolean ones]. Next, the class involved with PL includes expansions of Belnap's four-valued logic [3] (cf. [26]) by either implication or both bilattice connectives and classical negation (cf. [28] and [43]), in which case we get implicativity of the varieties of, resp. *implicative De Morgan lattices {algebras}* (cf. [41]) and *Boolean bilattices* (in particular, *implicative (Boolean) bilattices*; cf. [28]). In addition, the logic HZ [15] (cf. [50]) also belongs to the class under consideration with PL, in which case we get the implicativity (in particular, REDPRC) in the quasivariety of HZ-algebras (cf. [35]), that is not a variety, for $\theta \in \mathrm{Con}(\mathfrak{H}_3)$, where $\langle 0, 1 \rangle \notin \theta$, whereas $1\theta2$, while $(\mathfrak{H}_3/\theta) \notin \mathsf{HZ}$, because the quasi-identity $\neg x \approx x \rightarrow x \lesssim y$ being satisfied in $\mathfrak{H}_3$ is not true in $\mathfrak{H}_3/\theta$ under the assignment $[x/[2]_\theta, y/[0]_\theta]$. Further, the class under consideration with PL covers all *proper* non-classical axiomatic extension of P^ω found and studied in [42] (including Sette's P^1 [47]), in which case we get REDPRC and the implicativity of respective *proper* quasivarieties (including the one of *Sette algebras*; cf. [25]). Next, the class under consideration equally covers RM an its normal extensions (cf. [7], [48], [49]). In that case, RM_3 functionally equivalent to the *logic of antinomies LA* [1] holds PL. Finally, Łukasiewicz's finitely-valued logics [18] also belong to the class involved with PL (cf. Section 2.7 of [43]).

We do not discuss such examples in detail. Instead, we concentrate on analyzing algebraizable sequent calculi with DT [and, in most cases but that studied in Section 3.3, PL]. And what is more, we restrict our consideration only by most representative and useful (for either Universal Algebra or General Sentential Logic) examples of such a kind. On the other hand, to demonstrate the extensive applicability of the generic approach elaborated in Section 2.6, we study *all known* contraposable enlargable sequent calculi with algebraizable extensions by the Weak Contraposition rule. Likewise, we cannot refrain from providing extensive applications of Theorem 2.23, demonstrating the power of this fundamental generic result. On the other hand, we have refrained from considering all the examples discussed in [28] but merely those which are contraposabe, non-algebraizable, and whose extensions by the Weak Contraposition rule are algebraizable, while studying the rest is straightforward.

The *semilattice identities for* a binary $\diamond \in F$ are the following three identities:

$$\begin{aligned}
(x \diamond x) &\approx x, \\
(x \diamond y) &\approx (y \diamond x), \\
((x \diamond y) \diamond z) &\approx (x \diamond (y \diamond z)).
\end{aligned}$$

As usual, in case the functional signature F under consideration contains $\wedge$, we put $(x \lessapprox y) \triangleq ((x \wedge y) \approx x)$. In that case, given an F-algebra $\mathfrak{A}$ satisfying semilattice identities for $\wedge$, the partial ordering on A determined by $\wedge^{\mathfrak{A}}$ is denoted by $\leqslant^{\mathfrak{A}}$, that is, $a \leqslant^{\mathfrak{A}} b \Leftrightarrow a = a \wedge^{\mathfrak{A}} b$, for all $a, b \in A$.

Let $\mathfrak{A}$ be an F-algebra and $B \subseteq A$. Given a unary $\wr \in F$, B is said to be $\wr$(*-involutive*) [*-complementary*] in $\mathfrak{A}$, provided, for all $a \in A$, $a \in B \Leftrightarrow (\wr^{\mathfrak{A}} \wr^{\mathfrak{A}} a \in B)[\wr^{\mathfrak{A}}a \notin B]$. Given a binary $\diamond \in F$, B is (referred to as) [said to be] (a $\diamond$-*filter of*) [$\diamond$-*prime in*] $\mathfrak{A}$, provided, for all $a, b \in A$, $(a \diamond^{\mathfrak{A}} b) \in B \Leftrightarrow \{a, b\}(\subseteq B)[\cap B \neq \varnothing]$.

3.1 Distributive lattices and their expansions

Let $F \triangleq \{\wedge, \vee\}$. A *distributive lattice* (cf. [2], [12]) is any F-algebra satisfying the semilattice identities for both $\wedge$ and $\vee$ as well as the following two identities:[1]

$$\begin{aligned}
(x \wedge y) \vee x &\approx x, \\
(x \wedge (y \vee z)) &\approx ((x \wedge y) \vee (x \wedge z)).
\end{aligned}$$

---

[1]The identities dual to the mentioned ones are then derivable.

The variety of all distributive lattices is denoted by DL. Given any $n \in \omega$, by $\mathfrak{D}_n$ we denote the distributive lattice with carrier n given by the natural ordering on n.

Lemma 3.1. *Let $\mathfrak{B}$ be an F-algebra and X a $\wedge$-filter of $\mathfrak{B}$ being $\vee$-prime in it. Then, $\chi_B^X \in \mathrm{Hom}(\mathfrak{B}, \mathfrak{D}_2)$.*

Proof. Straightforward. $\square$

Corollary 3.2 ([2]). $\mathrm{DL} = \mathbf{I}(\mathfrak{D}_2)$.

Proof. The inclusion from right to left is trivial, for $\mathfrak{D}_2 \in \mathrm{DL}$. To prove the converse, consider any $\mathfrak{B} \in \mathrm{DL}$. Take arbitrary distinct $a, b \in B$. Then, $c \triangleq a \wedge^{\mathfrak{B}} b \not\leqslant^{\mathfrak{B}} d \triangleq a \vee^{\mathfrak{B}} b$. Hence, by the Prime Ideal Theorem for distributive lattices, there is a $\wedge$-filter X of $\mathfrak{B}$ being $\vee$-prime in it such that $c \in X \not\ni d$. Then, by Lemma 3.1, $\chi_B^X \in \mathrm{Hom}(\mathfrak{B}, \mathfrak{D}_2)$, while $\chi_B^X(c) = 1 \neq 0 = \chi_B^X(d)$, in which case $\chi_B^X(a) \neq \chi_B^X(b)$, as required. $\square$

The F-matrix $\mathcal{A} \triangleq \langle \mathfrak{D}_2, \{1\} \rangle$ with equality determinant $\Upsilon \triangleq \{x_0\}$ (cf. [38]) defines the positive constant-free fragment of the classical logic. Clearly, for each $i \in 2$, $\{i\} \subseteq S_i^A$ forms a subalgebra of $\mathfrak{A}$, in which case $\gamma_i^A = 1$. The F-sequent calculus $\mathbb{G}$ of type $[\omega \setminus 1, \omega \setminus 1]$, associated with $\mathcal{A}$ according to [38], can be chosen to consist exactly of classical and structural rules of the same type.

Though the following result is a particular case of (A.2), to demonstrate applicability and usefulness of Theorem 2.23, we give a direct (and quite transparent) proof of the result under consideration based upon the mentioned theorem.

Proposition 3.3. $\vdash_{\mathbb{G}} = \vdash_{\mathcal{A}^{[\omega \setminus 1, \omega \setminus 1]}}$.

Proof. First, by Theorem 2.23(ii)$\Rightarrow$(i), structural rules are true in $\mathfrak{A}^{[\omega \setminus 1, \omega \setminus 1]}$. Further, remark that D^A is a $\wedge$-filter of $\mathfrak{A}$ being $\vee$-prime in it. Therefore, the axioms (2.19)—(2.24) are true in $\mathcal{A}^{[\omega \setminus 1, \omega \setminus 1]}$. Hence, in view of Remark 2.31, classical rules are true in $\mathfrak{A}^{[\omega \setminus 1, \omega \setminus 1]}$. Thus, $\vdash_{\mathbb{G}} \subseteq \vdash_{\mathcal{A}^{[\omega \setminus 1, \omega \setminus 1]}}$. Conversely, by Theorem 2.23(i)$\Rightarrow$(ii), there is some class K of F-matrices such that $\vdash_{\mathbb{G}} = \vdash_{\mathsf{K}^{[\omega \setminus 1, \omega \setminus 1]}}$. Consider any $\mathcal{B} \in \mathsf{K}$. Then, in view of Remark 2.31, the axioms (2.19)—(2.24) are true in $\mathcal{B}^{[\omega \setminus 1, \omega \setminus 1]}$. Hence, $D^{\mathcal{B}}$ is a $\wedge$-filter of $\mathfrak{B}$ being $\vee$-prime in it. Therefore, by Lemma 3.1, $\chi_B^{D^{\mathcal{B}}} \in \mathrm{SHom}(\mathcal{B}, \mathcal{A})$, and so $\chi_B^{D^{\mathcal{B}}} \in \mathrm{SHom}(\mathcal{B}^{[\omega \setminus 1, \omega \setminus 1]}, \mathcal{A}^{[\omega \setminus 1, \omega \setminus 1]})$, in view of Remark 2.22. Then, by Lemma 1.14 of [42], $\vdash_{\mathcal{A}^{[\omega \setminus 1, \omega \setminus 1]}} \subseteq \vdash_{\mathcal{B}^{[\omega \setminus 1, \omega \setminus 1]}}$. Thus, $\vdash_{\mathcal{A}^{[\omega \setminus 1, \omega \setminus 1]}} \subseteq \vdash_{\mathbb{G}}$, as required. $\square$

Define the purely relational parameter-less $(F^{[\omega\setminus 1,\omega\setminus 1]}, F + \approx)$-translation δ as follows: for all $m, n \in \omega \setminus 1$, put:

$$\delta(\mapsto{}^m_n) \triangleq \{\wedge\overline{x}(0, m) \lesssim \vee\overline{x}(m, n)\}.$$

Since $D^{\mathcal{A}}$ is a $\wedge$-filter of $\mathfrak{D}_2$ being $\vee$-prime in it, while $(a = 1 \Rightarrow b = 1) \Leftrightarrow a \leqslant b$, for all $a, b \in 2$, it is easy to see that:

$$(3.1) \qquad\qquad \mathcal{A}^{[\omega\setminus 1,\omega\setminus 1]} = \delta^{-1}[\mathfrak{D}_2 + \approx].$$

In view of Theorem 2.10(v)$\Rightarrow$(i) of [42], Proposition 3.3, Corollary 3.2, (3.1) and (A.1) immediately yield:

Proposition 3.4 (cf. [27], [39], [43]). *Both $\mathfrak{D}_2$ and DL are finitary relationally equivalent purely-algebraic semantics for $\mathbb{G}$ with respect to $\varepsilon_{\{x_0\}}$ and δ.*

Consider the binary $(F + \approx)$-system λ_{DL}, constituted by the following four F-equations over V^4:

$$(3.2) \qquad\qquad x_2 \lesssim x_0 \vee x_1 \vee x_3,$$
$$(3.3) \qquad\qquad x_0 \wedge x_1 \wedge x_2 \lesssim x_3,$$
$$(3.4) \qquad\qquad x_3 \lesssim x_0 \vee x_1 \vee x_2,$$
$$(3.5) \qquad\qquad x_0 \wedge x_1 \wedge x_3 \lesssim x_2.$$

Using Theorem 2.8, Proposition 3.4, Corollary 2.17 together with (2.8) and omitting tautological (in distributive lattices) equations (viz., identities true in DL) of the form $\delta(\overline{\varphi} \mapsto \overline{\psi})$, where $(\overline{\varphi} \mapsto \overline{\psi}) \in \mathrm{Seq}_\varnothing^{[\omega\setminus 1,\omega\setminus 1]}(V^4)$, while $(\mathrm{img}\,\overline{\varphi}) \cap (\mathrm{img}\,\overline{\psi}) \neq \varnothing$, by Theorem 2.19, we eventually obtain:

Corollary 3.5 (cf. [14] for REDPC alone). *DL is λ_{DL}-implicative.*

Since expansions by constants respect congruences, while two-element algebras are algebraically simple, any expansions of distributive lattices by constants alone inherit the above implicativity schema, whenever their prevarieties are generated by corresponding expansions of $\mathfrak{D}_2$. In particular, *[partially] bounded* distributive lattices (viz., those with zero and[/or] unit) are λ_{DL}-implicative. Moreover, *distributive bilattices* [10] are rationally equivalent to expansions of distributive lattices by certain constants with respect to an (F', F)-translation and ι_F (cf. [31]), while the variety of formers is the prevariety generated by expansions of $\mathfrak{D}_2$ (cf. [31]). Hence, by Corollaries 2.16 and 3.5, the variety of distributive bilattices is λ_{DL}-implicative as well, for

$\iota_F[\lambda_{\mathsf{DL}}] = \lambda_{\mathsf{DL}}$. Finally, since congruences of Boolean algebras are exactly those of their F-reducts (cf., e.g., the proof of Lemma 3.1 of [44]), both the variety of *Boolean algebras* (cf. [2]) and that of *Boolean bilattices* (viz., expansions of Boolean algebras by bilattice connectives; cf. [28] and [31]) are equally λ_{DL}-implicative, for they are the prevarieties generated by expansions of $\mathfrak{D}_2$.

3.1.1 Bounded distributive lattices

Let $F_{01} \triangleq F \cup \{\bot, \top\}$. A *bounded distributive lattice* (cf. [2]) is any F_{01}-algebra satisfying the identities axiomatizing the variety DL together with the following additional identities:

$$(3.6) \qquad\qquad x \lesssim \gtrsim \top,$$

$$(3.7) \qquad\qquad \bot \lesssim \gtrsim x.$$

The variety of all bounded distributive lattices is denoted by BDL. Given any $n \in \omega$, by $\mathfrak{BD}_n$ we denote the bounded distributive lattice with carrier n given by the natural ordering on n.

Lemma 3.6. *Let $\mathfrak{B}$ be an F_{01}-algebra and X a $\wedge$-filter of $\mathfrak{B}$ being $\vee$-prime in it and containing $\top^{\mathfrak{B}}$ but not containing $\bot^{\mathfrak{B}}$. Then, $\chi_B^X \in \mathrm{Hom}(\mathfrak{B}, \mathfrak{BD}_2)$.*

Proof. By Lemma 3.1. □

Corollary 3.7 ([2]). $\mathrm{BDL} = \mathrm{I}(\mathfrak{BD}_2)$.

Proof. The inclusion from right to left is trivial, for $\mathfrak{BD}_2 \in \mathrm{BDL}$. To prove the converse, consider any $\mathfrak{B} \in \mathrm{BDL}$. Take arbitrary distinct $a, b \in B$. Then, $c \triangleq a \wedge^{\mathfrak{B}} b \not\leq^{\mathfrak{B}} d \triangleq a \vee^{\mathfrak{B}} b$. Hence, by the Prime Ideal Theorem for distributive lattices, there is a $\wedge$-filter X of $\mathfrak{B}$ being $\vee$-prime in it such that $c \in X \not\ni d$, in which case $\top^{\mathfrak{B}} \in X \not\ni \bot^{\mathfrak{B}}$, by (3.6), (3.7). Then, by Lemma 3.6, $\chi_B^X \in \mathrm{Hom}(\mathfrak{B}, \mathfrak{BD}_2)$, while $\chi_B^X(c) = 1 \neq 0 = \chi_B^X(d)$, in which case $\chi_B^X(a) \neq \chi_B^X(b)$, as required. □

The F_{01}-matrix $\mathcal{A} \triangleq \langle \mathfrak{BD}_2, \{1\} \rangle$ with the same equality determinant Υ (see above) defines the full positive fragment of the classical logic. Clearly, for each $i \in 2$, $\gamma_i^{\mathcal{A}} = 0$. The F_{01}-sequent calculus $\mathbb{G}$ of type $[\omega, \omega]$, associated with $\mathcal{A}$ according to [38], can be chosen to consist exactly of classical and structural rules of the same type together with the axioms:

$$(3.8) \qquad\qquad \longmapsto \top,$$

$$(3.9) \qquad\qquad \bot \longmapsto .$$

It is the full positive propositional fragment of Gentzen's LK [9].

Though the following result is a particular case of (A.2), to demonstrate applicability and usefulness of Theorem 2.23, we give a direct (and quite transparent) proof of the result under consideration based upon the mentioned theorem.

Proposition 3.8. $\vdash_{\mathbb{G}} = \vdash_{\mathcal{A}^{[\omega,\omega]}}$.

Proof. First, by Theorem 2.23(ii)$\Rightarrow$(i), structural rules are true in $\mathfrak{A}^{[\omega,\omega]}$. Further, remark that $D^{\mathcal{A}}$ is a $\wedge$-filter of $\mathfrak{A}$ being $\vee$-prime in it and containing $\top^{\mathfrak{A}}$ but not containing $\bot^{\mathfrak{A}}$. Therefore, the axioms (2.19)—(2.24), (3.8) and (3.9) are true in $\mathcal{A}^{[\omega,\omega]}$. Hence, in view of Remark 2.31, classical rules are true in $\mathfrak{A}^{[\omega,\omega]}$. Thus, $\vdash_{\mathbb{G}} \subseteq \vdash_{\mathcal{A}^{[\omega,\omega]}}$. Conversely, by Theorem 2.23(i)$\Rightarrow$(ii), there is some class K of F_{01}-matrices such that $\vdash_{\mathbb{G}} = \vdash_{K^{[\omega,\omega]}}$. Consider any $\mathcal{B} \in$ K. Then, in view of Remark 2.31, the axioms (2.19)—(2.24), (3.8) and (3.9) are true in $\mathcal{B}^{[\omega,\omega]}$. Hence, $D^{\mathcal{B}}$ is a $\wedge$-filter of $\mathfrak{B}$ being $\vee$-prime in it and containing $\top^{\mathfrak{B}}$ but not containing $\bot^{\mathfrak{B}}$. Therefore, by Lemma 3.6, $\chi_B^{D^{\mathcal{B}}} \in \mathrm{SHom}(\mathcal{B}, \mathcal{A})$, and so $\chi_B^{D^{\mathcal{B}}} \in \mathrm{SHom}(\mathcal{B}^{[\omega,\omega]}, \mathcal{A}^{[\omega,\omega]})$, in view of Remark 2.22. Then, by Lemma 1.14 of [42], $\vdash_{\mathcal{A}^{[\omega,\omega]}} \subseteq \vdash_{\mathcal{B}^{[\omega,\omega]}}$. Thus, $\vdash_{\mathcal{A}^{[\omega,\omega]}} \subseteq \vdash_{\mathbb{G}}$, as required. $\square$

Define the purely relational parameter-less $(F_{01}^{[\omega,\omega]}, F_{01} + \approx)$-translation δ_{01} as follows: for all $m, n \in \omega$, put:

$$\delta_{01}(\rightarrowtail_n^m) \triangleq \{\wedge(\top, \bar{x}(0, m)) \lesssim \vee(\bot, \bar{x}(m, n))\}.$$

Since $D^{\mathcal{A}}$ is a $\wedge$-filter of $\mathfrak{B}\mathfrak{D}_2$ being $\vee$-prime in it and containing $\top^{\mathfrak{A}}$ but not containing $\bot^{\mathfrak{A}}$, while $(a = 1 \Rightarrow b = 1) \Leftrightarrow a \leqslant b$, for all $a, b \in 2$, it is easy to see that:

(3.10) $$\mathcal{A}^{[\omega,\omega]} = (\delta_{01})^{-1}[\mathfrak{B}\mathfrak{D}_2 + \approx].$$

In view of Theorem 2.10(v)$\Rightarrow$(i) of [42], Proposition 3.8, Corollary 3.7, (3.10) and (A.1) immediately yield:

Proposition 3.9 (cf. [39], [43]). *Both $\mathfrak{B}\mathfrak{D}_2$ and BDL are finitary relationally equivalent purely-algebraic semantics for $\mathbb{G}$ with respect to $\varepsilon_{\{x_0\}}$ and δ_{01}.*

3.1.2   Boolean algebras

Let $F_{\mathrm{B}} \triangleq F_{01} \cup \{\neg\}$, where $\neg$ is unary (*negation*). We also use the following secondary binary connectives:

$$x_0 \supset x_1 \triangleq (\neg x_0 \vee x_1),$$
$$x_0 \leftrightarrows x_1 \triangleq (x_0 \supset x_1) \wedge (x_1 \supset x_0).$$

A *Boolean algebra* (cf. [2]) is any F_B-algebra satisfying the identities axiomatizing the variety BDL together with the following two additional identities:

$$(3.11) \qquad\qquad x \vee \neg x \approx \top,$$
$$(3.12) \qquad\qquad x \wedge \neg x \approx \bot.$$

The variety of all Boolean algebras is denoted by BA. By $\mathfrak{B}_2$ we denote the Boolean algebra with carrier 2 given by $0 < 1$.

Lemma 3.10. *Let $\mathfrak{B}$ be an F_B-algebra and X a $\wedge$-filter of $\mathfrak{B}$ being both $\neg$-complementary and $\vee$-prime in it and containing $\top^{\mathfrak{B}}$ but not containing $\bot^{\mathfrak{B}}$. Then, $\chi_B^X \in \mathrm{Hom}(\mathfrak{B}, \mathfrak{B}_2)$.*

Proof. By Lemma 3.6. $\qquad\qquad\qquad\qquad\qquad\qquad\qquad\qquad\qquad\qquad\qquad\square$

Corollary 3.11 ([2]). $\mathrm{BA} = \mathbf{I}(\mathfrak{B}_2)$.

Proof. The inclusion from right to left is trivial, for $\mathfrak{B}_2 \in \mathrm{BA}$. To prove the converse, consider any $\mathfrak{B} \in \mathrm{BA}$. Take arbitrary distinct $a, b \in B$. Then, $c \triangleq a \wedge^{\mathfrak{B}} b \not\leq^{\mathfrak{B}} d \triangleq a \vee^{\mathfrak{B}} b$. Hence, by the Prime Ideal Theorem for distributive lattices, there is a $\wedge$-filter X of $\mathfrak{B}$ being $\vee$-prime in it such that $c \in X \not\ni d$, in which case $\top^{\mathfrak{B}} \in X \not\ni \bot^{\mathfrak{B}}$, by (3.6), (3.7). Then, by (3.11) and (3.12), X is $\neg$-complementary in $\mathfrak{B}$. Hence, by Lemma 3.6, $\chi_B^X \in \mathrm{Hom}(\mathfrak{B}, \mathfrak{B}_2)$, while $\chi_B^X(c) = 1 \neq 0 = \chi_B^X(d)$, in which case $\chi_B^X(a) \neq \chi_B^X(b)$, as required. $\qquad\square$

The F_B-matrix $\mathcal{A} \triangleq \langle \mathfrak{B}_2, \{1\} \rangle$ with the same equality determinant Υ (see above) defines the classical logic $\mathbb{PC}$. Clearly, for each $i \in 2$, $\gamma_i^{\mathcal{A}} = 0$. The F_B-sequent calculus $\mathbb{G}$ of type $[\omega, \omega]$, associated with $\mathcal{A}$ according to [38], can be chosen to consist exactly of classical and structural rules of the same type together with the axioms (3.8), (3.9) and the following additional rules (where $m, n \in \omega$):

$$(3.13) \qquad \frac{\overline{x}(1, m) \rightarrowtail \overline{x}(m+1, n), x_0}{\neg x_0, \overline{x}(1, m) \rightarrowtail \overline{x}(m+1, n)},$$

$$(3.14) \qquad \frac{x_0, \overline{x}(m+1, n) \rightarrowtail \overline{x}(1, m)}{\overline{x}(m+1, n) \rightarrowtail \neg x_0, \overline{x}(1, m)}.$$

This is the propositional fragment of Gentzen's LK [9].

Remark 3.12. Under presence of structural rules, (3.13) and (3.14) are reversible. More precisely, the following axioms are immediately derivable from Reflexivity by means of (3.13) and (3.14) of type $[\omega, \omega]$, respectively:

$$(3.15) \qquad\qquad \neg x_0, x_0 \ \longmapsto \ ,$$

$$(3.16) \qquad\qquad\qquad\qquad \longmapsto \ \neg x_0, x_0.$$

On the other hand, the rules inverse to (3.13) and (3.14) of the same type (as well as the latter themselves) are derivable from the axioms (3.15) and (3.16) by means of structural rules of the same type. Moreover, these axioms are equally derivable from Reflexivity by means of the rules inverse to (3.13) and (3.14) of the same type. In particular, the rules (3.13) and (3.14) of the same type and those which are inverse to them are, so to say, inter-derivable modulo structural rules of the same type. □

Though the following result is a particular case of (A.2), to demonstrate applicability and usefulness of Theorem 2.23, we give a direct (and quite transparent) proof of the result under consideration based upon the mentioned theorem.

Proposition 3.13. $\vdash_{\mathbb{G}} = \vdash_{\mathcal{A}^{[\omega,\omega]}}$.

Proof. First, by Theorem 2.23(ii)⇒(i), structural rules are true in $\mathfrak{A}^{[\omega,\omega]}$. Further, remark that D^A is a $\wedge$-filter of $\mathfrak{A}$ being both $\neg$-complementary and $\vee$-prime in it and containing $\top^{\mathfrak{A}}$ but not containing $\bot^{\mathfrak{A}}$. Therefore, the axioms (2.19)—(2.24), (3.8), (3.9), (3.15) and (3.16) are true in $\mathcal{A}^{[\omega,\omega]}$. Hence, in view of Remarks 2.31 and 3.12, classical rules as well the rules (3.13) and (3.14) are true in $\mathfrak{A}^{[\omega,\omega]}$. Thus, $\vdash_{\mathbb{G}} \subseteq \vdash_{\mathcal{A}^{[\omega,\omega]}}$. Conversely, by Theorem 2.23(i)⇒(ii), there is some class K of F_{B}-matrices such that $\vdash_{\mathbb{G}} = \vdash_{\mathsf{K}^{[\omega,\omega]}}$. Consider any $\mathcal{B} \in \mathsf{K}$. Then, in view of Remarks 2.31 and 3.12, the axioms (2.19)—(2.24), (3.8), (3.9), (3.15) and (3.16) are true in $\mathcal{B}^{[\omega,\omega]}$. Hence, $D^{\mathcal{B}}$ is a $\wedge$-filter of $\mathfrak{B}$ being both $\neg$-complementary and $\vee$-prime in it and containing $\top^{\mathfrak{B}}$ but not containing $\bot^{\mathfrak{B}}$. Therefore, in view of Lemma 3.10, $\chi_B^{D^{\mathcal{B}}} \in \mathrm{SHom}(\mathcal{B}, \mathcal{A})$, and so $\chi_B^{D^{\mathcal{B}}} \in \mathrm{SHom}(\mathcal{B}^{[\omega,\omega]}, \mathcal{A}^{[\omega,\omega]})$, in view of Remark 2.22. Then, by Lemma 1.14 of [42], $\vdash_{\mathcal{A}^{[\omega,\omega]}} \subseteq \vdash_{\mathcal{B}^{[\omega,\omega]}}$. Thus, $\vdash_{\mathcal{A}^{[\omega,\omega]}} \subseteq \vdash_{\mathbb{G}}$, as required. □

In view of Theorem 2.10(v)⇒(i) of [42], Proposition 3.13, Corollary 3.11, (3.10) and (A.1) immediately yield:

Proposition 3.14 (cf. [27], [39], [43]). *Both $\mathfrak{B}_2$ and* BA *are finitary relationally equivalent purely-algebraic semantics for* $\mathbb{G}$ *with respect to* $\varepsilon_{\{x_0\}}$ *and* δ_{01}.

Consider the purely relational parameter-less $(\langle F_{\mathrm{B}}, \{D\}\rangle F_{\mathrm{B}}^{[\omega,\omega]})$-translation ϱ, given by $\varrho(D) \triangleq \{\longmapsto x_0\}$, and $(\langle F_{\mathrm{B}}^{[\omega,\omega]}, F_{\mathrm{B}}\rangle\{D\})$-translation ϑ, defined by (where $m, n \in \omega$):

$$\vartheta(\longmapsto_n^m) \triangleq \{\vee(\neg\overline{x}(0,m), \overline{x}(m,n), \bot)\}.$$

Since $D^{\mathcal{A}} \not\ni \bot^{\mathfrak{A}}$ is both $\vee$-prime and $\neg$-complementary in $\mathfrak{A}$, it is easy to see that:

$$(3.17) \qquad \mathcal{A}^{[\omega,\omega]} = \vartheta^{-1}[\mathcal{A}],$$
$$(3.18) \qquad \mathcal{A} = \varrho^{-1}[\mathcal{A}^{[\omega,\omega]}].$$

Theorem 2.10(v)$\Rightarrow$(i) of [42] together with Proposition 3.13, (3.17) and (3.18) immediately yield:

Proposition 3.15 (cf. [27], [43]). $\mathbb{PC}$ *and* $\mathbb{G}$ *are relationally equivalent with respect to* ϱ *and* ϑ.

3.2 De Morgan lattices and their expansions

Throughout the rest of this chapter, we adopt the following conventions. Put $\widetilde{F}_{[k]} \triangleq F_{[k]} \cup \{\sim\}[$, where k is either "01" or "B"]. Given an $\widetilde{F}_{[k]}$-algebra $\mathfrak{B}$ and $X \subseteq B$, put $h_{\langle\mathfrak{B},X\rangle} \triangleq h_{\mathfrak{B}}^X : B \to 2^2, a \mapsto \langle\chi_B^X(a), \chi_B^{B\setminus(\sim^{\mathfrak{B}})^{-1}[X]}(a)\rangle$.

Recall that a *De Morgan lattice* (cf. [2], [16], [21], [30]) is any $\widetilde{F}$-algebra, whose F-reduct is a distributive lattice and the following identities are true in it:

$$(3.19) \qquad \sim\sim x \approx x,$$
$$(3.20) \qquad \sim(x \vee y) \approx \sim x \wedge \sim y,$$
$$(3.21) \qquad \sim(x \wedge y) \approx \sim x \vee \sim y,$$

the variety of of all them being denoted by DML.

By $\mathfrak{DM}_4$ we denote the diamond De Morgan lattice, whose F-reduct is $(\mathfrak{D}_2)^2$ and whose weak negation is given by $\sim^{\mathfrak{DM}_4}\langle a, b\rangle \triangleq \langle 1-b, 1-a\rangle$, for all $a, b \in 2$.

Lemma 3.16. *Let* $\mathfrak{B}$ *be an* $\widetilde{F}$-*algebra and* X *a* $\wedge$-*filter of* $\mathfrak{B}$ *being both* $\sim$-*involutive and* $\vee$-*prime in it. Suppose* $(\sim^{\mathfrak{B}})^{-1}[X]$ *is a* $\vee$-*filter of* $\mathfrak{B}$ *being* $\wedge$-*prime in it. Then,* $h_{\mathfrak{B}}^X \in \mathrm{Hom}(\mathfrak{B}, \mathfrak{DM}_4)$.

44

Proof. In that case, $B \setminus (\sim^{\mathfrak{B}})^{-1}[X]$ is a $\wedge$-filter of $\mathfrak{B}$ being $\vee$-prime in it. Hence, by Lemma 3.1, $h_{\mathfrak{B}}^X \in \mathrm{Hom}(\mathfrak{B} \upharpoonright F, (\mathfrak{D}_2)^2)$. The rest of argumentation is straightforward with using the $\sim$-involutivity of X in $\mathfrak{B}$. □

Corollary 3.17 ([16], [30]). DML $= \mathbf{I}(\mathfrak{D}\mathfrak{M}_4)$.

Proof. The inclusion from right to left is trivial, for $\mathfrak{D}\mathfrak{M}_4 \in$ DML. To prove the converse, consider any $\mathfrak{B} \in$ DML. Take arbitrary distinct $a, b \in B$. Then, $c \triangleq a \wedge^{\mathfrak{B}} b \not\leqslant^{\mathfrak{B}} d \triangleq a \vee^{\mathfrak{B}} b$. Hence, by the Prime Ideal Theorem for distributive lattices, there is a $\wedge$-filter X of $\mathfrak{B}$ $\vee$-prime in it such that $c \in X \not\ni d$. Moreover, by (3.19), X is $\sim$-involutive in $\mathfrak{B}$. And what is more, by (3.20) and (3.21), $(\sim^{\mathfrak{B}})^{-1}[X]$ is a $\vee$-filter of $\mathfrak{B}$ being $\wedge$-prime in it. Then, by Lemma 3.16, $h_{\mathfrak{B}}^X \in \mathrm{Hom}(\mathfrak{B}, \mathfrak{D}\mathfrak{M}_4)$, while $\pi_0 h_{\mathfrak{B}}^X(c) = 1 \neq 0 = \pi_0 h_{\mathfrak{B}}^X(d)$, in which case $h_{\mathfrak{B}}^X(a) \neq h_{\mathfrak{B}}^X(b)$, as required. □

The $\widetilde{F}$-matrix $\mathcal{A} \triangleq \langle \mathfrak{D}\mathfrak{M}_4, (\pi_0)^{-1}[\{1\}] \rangle$ with equality determinant

$$\widetilde{\Upsilon} \triangleq \{x_0, \sim x_0\}$$

(cf. [38]) defines Belnap's four-valued logic [3] (cf. [26]). Clearly, for each $i \in 2$, $\{\langle i, 1-i \rangle\} \subseteq S_i^{\mathcal{A}}$ forms a subalgebra of $\mathfrak{A}$, in which case $\gamma_i^{\mathcal{A}} = 1$. The $\widetilde{F}$-sequent calculus $\mathbb{G}$ of type $[\omega \setminus 1, \omega \setminus 1]$, associated with $\mathcal{A}$ according to [38], can be chosen to consist exactly of classical, De Morgan, involution and structural rules of the same type. This is exactly the both implication-less and $[\omega \setminus 1, \omega \setminus 1]$-fragment of Popov's calculus $GPar$ [23]. It was studied in [26], [27], [28] and [43]. According to the latter three works, $\mathbb{G}$ is not algabraizable (cf. [39]).

Though the following result is a particular case of (A.2), to demonstrate applicability and usefulness of Theorem 2.23, we give a direct (and quite transparent) proof of the result under consideration based upon the mentioned theorem.

Proposition 3.18. $\vdash_{\mathbb{G}} = \vdash_{\mathcal{A}^{[\omega \setminus 1, \omega \setminus 1]}}$.

Proof. First, by Theorem 2.23(ii)⇒(i), structural rules are true in $\mathfrak{A}^{[\omega \setminus 1, \omega \setminus 1]}$. Further, remark that $D^{\mathcal{A}}$ is a $\wedge$-filter of $\mathfrak{A}$ being both $\vee$-prime and $\sim$-involutive in it, while $(\sim^{\mathfrak{A}})^{-1}[D^{\mathcal{A}}]$ is a $\vee$-filter of $\mathfrak{A}$ being $\wedge$-prime in it. Therefore, the axioms (2.19)—(2.30), (2.17) and (2.18) are true in $\mathcal{A}^{[\omega \setminus 1, \omega \setminus 1]}$. Hence, in view of Remarks 2.28 and 2.31, classical, De Morgan and involution rules are true in $\mathfrak{A}^{[\omega \setminus 1, \omega \setminus 1]}$. Thus, $\vdash_{\mathbb{G}} \subseteq \vdash_{\mathcal{A}^{[\omega \setminus 1, \omega \setminus 1]}}$. Conversely, by Theorem 2.23(i)⇒(ii), there is some class K of $\widetilde{F}$-matrices such that $\vdash_{\mathbb{G}} = \vdash_{\mathrm{K}^{[\omega \setminus 1, \omega \setminus 1]}}$. Consider any $\mathcal{B} \in$ K. Then, in view of Remarks 2.28 and 2.31, the axioms (2.19)—(2.30), (2.17)

45

and (2.18) are true in $\mathcal{B}^{[\omega\backslash 1,\omega\backslash 1]}$. Hence, $D^{\mathcal{B}}$ is a $\wedge$-filter of $\mathfrak{B}$ being both $\sim$-involutive and $\vee$-prime in it, while $(\sim^{\mathfrak{B}})^{-1}[D^{\mathcal{B}}]$ is a $\vee$-filter of $\mathfrak{A}$ being $\wedge$-prime in it. Therefore, in view of Lemma 3.16, we have $h_{\mathcal{B}} \in \mathrm{SHom}(\mathcal{B}, \mathcal{A})$, and so $h_{\mathcal{B}} \in \mathrm{SHom}(\mathcal{B}^{[\omega\backslash 1,\omega\backslash 1]}, \mathcal{A}^{[\omega\backslash 1,\omega\backslash 1]})$, in view of Remark 2.22. Then, by Lemma 1.14 of [42], $\vdash_{\mathcal{A}^{[\omega\backslash 1,\omega\backslash 1]}} \subseteq \vdash_{\mathcal{B}^{[\omega\backslash 1,\omega\backslash 1]}}$. Thus, $\vdash_{\mathcal{A}^{[\omega\backslash 1,\omega\backslash 1]}} \subseteq \vdash_{\mathbb{G}}$, as required. $\qquad\square$

Lemma 3.19. $\mathbb{G}$ *is weakly contraposable.*

Proof. First, (2.11) immediately follows from (2.15) and (2.16). The rest of argumentation is straightforward with using (2.15), (2.16) and Remark 2.28. $\qquad\square$

Clearly, $\mathbb{G}$ is normal. Then, by Corollary 2.33 and Lemma 3.19, we immediately obtain:

Theorem 3.20. $\widehat{\mathbb{G}}$ *has DT and holds PL with respect to* $\lambda_{\widetilde{F}^{[\omega\backslash 1,\omega\backslash 1]}}$.

The following result was first proved in [27] (cf. [28], [43]) with using the proof-theoretical characterization of equivalence provided by Proposition 2.4 of [42] (extending an appropriate result of [27] itself). On the other hand, it equally follows from Lemma 2.10, Theorem 3.15 and Proposition 3.24 of [44] together with Proposition 3.18, Corollary 3.17 and the fact that $\{1\}$ is a $\wedge$-filter of $\mathfrak{D}_2$ being $\vee$-prime in it (this provides a much more transparent algebraic argumentation of the result under consideration).

Proposition 3.21. *Both* $\mathfrak{DM}_4$ *and* DML *are finitary relationally equivalent purely-algebraic semantics for* $\widehat{\mathbb{G}}$ *with respect to* $\varepsilon_{\{x_0\}}$ *and* δ.

Consider the binary $(\widetilde{F}+\approx)$-system λ_{DML}, constituted by the following eight $\widetilde{F}$-equations over V^4:

$$(3.22) \qquad\qquad x_2 \lesssim\!\!\!\!\!\gtrsim x_0 \vee x_1 \vee x_3 \vee \sim x_0 \vee \sim x_1,$$

$$(3.23) \qquad \sim x_0 \wedge \sim x_1 \wedge x_2 \lesssim\!\!\!\!\!\gtrsim x_0 \vee x_1 \vee x_3,$$

$$(3.24) \qquad x_0 \wedge x_1 \wedge x_2 \lesssim\!\!\!\!\!\gtrsim x_3 \vee \sim x_0 \vee \sim x_1,$$

$$(3.25) \quad \sim x_0 \wedge \sim x_1 \wedge x_0 \wedge x_1 \wedge x_2 \lesssim\!\!\!\!\!\gtrsim x_3,$$

$$(3.26) \qquad\qquad x_3 \lesssim\!\!\!\!\!\gtrsim x_0 \vee x_1 \vee x_2 \vee \sim x_0 \vee \sim x_1,$$

$$(3.27) \qquad \sim x_0 \wedge \sim x_1 \wedge x_3 \lesssim\!\!\!\!\!\gtrsim x_0 \vee x_1 \vee x_2,$$

$$(3.28) \qquad x_0 \wedge x_1 \wedge x_3 \lesssim\!\!\!\!\!\gtrsim x_2 \vee \sim x_0 \vee \sim x_1,$$

$$(3.29) \quad \sim x_0 \wedge \sim x_1 \wedge x_0 \wedge x_1 \wedge x_3 \lesssim\!\!\!\!\!\gtrsim x_2.$$

Put $\sigma \triangleq [x_j/\sim x_j]_{j\in(4\backslash 2)}$.

46

Lemma 3.22. $\lambda_{\mathsf{DML}} \vdash^{\approx}_{\mathsf{DML}} \sigma[\lambda_{\mathsf{DML}}]$.

Proof. Given any $\phi \in \mathrm{Tm}_{\sim}(V^4)$, put:

$$-\phi \triangleq \begin{cases} \psi & \text{if } \phi = {\sim}\psi, \text{ where } \psi \in \mathrm{Tm}_{\sim}(V^4), \\ {\sim}\phi & \text{otherwise}, \end{cases}$$

in which case $-\phi \in \mathrm{Tm}_{\sim}(V^4)$ as well, while, by (3.19), the following identity is true in DML:

$$(3.30) \qquad\qquad -\phi \approx {\sim}\phi.$$

Further, given any $n \in \omega$ and any $\overline{\phi} \in \mathrm{Tm}_{\sim}(V^4)^n$, set $-\overline{\phi} \triangleq \langle -\phi_i \rangle_{i \in n} \in \mathrm{Tm}_{\sim}(V^4)^n$. Finally, given any $\Phi = (\overline{\varphi} \longmapsto \overline{\psi}) \in \mathrm{Seq}_{\sim}^{[\omega \backslash 1, \omega \backslash 1]}(V^4)$, put $\overline{\Phi} \triangleq (-\overline{\psi} \longmapsto -\overline{\varphi}) \in \mathrm{Seq}_{\sim}^{[\omega \backslash 1, \omega \backslash 1]}(V^4)$. Notice that the mapping $\delta \upharpoonright \mathrm{Seq}_{\sim}^{[\omega \backslash 1, \omega \backslash 1]}(V^4)$ is injective. This makes the following definition correct:

$$\overline{\delta(\Phi)} \triangleq \delta(\overline{\Phi}),$$

where $\Phi \in \mathrm{Seq}_{\sim}^{[\omega \backslash 1, \omega \backslash 1]}(V^4)$. By (3.30), (3.20), (3.21), the following holds:

$$(3.31) \qquad\qquad \Psi \vdash^{\approx}_{\mathsf{DML}} \overline{\Psi},$$

where $\Psi \in \delta[\mathrm{Seq}_{\sim}^{[\omega \backslash 1, \omega \backslash 1]}(V^4)]$. Next, given any $\Gamma \subseteq \delta[\mathrm{Seq}_{\sim}^{[\omega \backslash 1, \omega \backslash 1]}(V^4)]$, put $\overline{\Gamma} \triangleq \{\overline{\Phi} | \Phi \in \Gamma\}$. It is routine checking that:

$$\begin{aligned} \sigma(3.22) &= \overline{(3.29)}, \\ \sigma(3.23) &= \overline{(3.27)}, \\ \sigma(3.24) &= \overline{(3.28)}, \\ \sigma(3.25) &= \overline{(3.26)}, \\ \sigma(3.26) &= \overline{(3.25)}, \\ \sigma(3.27) &= \overline{(3.23)}, \\ \sigma(3.28) &= \overline{(3.24)}, \\ \sigma(3.29) &= \overline{(3.22)}. \end{aligned}$$

Thus, $\sigma[\lambda_{\mathsf{DML}}] = \overline{\lambda_{\mathsf{DML}}}$. Then, (3.31) completes the argument. $\square$

Corollary 3.23 (cf. [46] for REDPC alone). DML *is* λ_{DML}-*implicative*.

Proof. Using Theorem 2.8, Proposition 3.21, Corollary 2.17, (2.14) and omitting tautological (in De Morgan lattices) equations (viz., identities true in DML) of the form $\delta(\overline{\varphi} \longmapsto \overline{\psi})$, where $(\overline{\varphi} \longmapsto \overline{\psi}) \in \operatorname{Seq}_{\sim}^{[\omega \backslash 1, \omega \backslash 1]}(V^4)$, while $(\operatorname{img} \overline{\varphi}) \cap (\operatorname{img} \overline{\psi}) \neq \varnothing$, by Theorem 3.20, we eventually conclude that DML is $(\lambda_{\mathsf{DL}} \cup \sigma[\lambda_{\mathsf{DL}}])$-implicative. Then, Lemma 3.22 completes the argument. $\square$

Since expansions by constants respect congruences (in particular, algebraic simplicity), any expansions of De Morgan lattices by constants alone inherit the above implicativity schema, provided prevarieties of them are generated by corresponding expansions of $\mathfrak{DM}_4$. In particular, the variety of *De Morgan algebras* (viz., De Morgan lattices being also bounded distributive lattices; cf. [2]) is λ_{DML}-implicative. Moreover, *distributive bilattices with negation* (viz, bilattice expansions of De Morgan algebras; cf. [10], [28], [31]) are rationally equivalent to expansions of De Morgan lattices by certain constants with respect to an $(F', \widetilde{F})$-translation and $\iota_{\widetilde{F}}$ (cf. [31]). Hence, by Corollaries 2.16 and 3.23, the variety of formers is λ_{DML}-implicative as well, for $\iota_{\widetilde{F}}[\lambda_{\mathsf{DML}}] = \lambda_{\mathsf{DML}}$. Finally, since congruences of Boolean algebras are exactly those of their lattice reducts (cf., e.g., the proof of Lemma 3.1 of [44]), both the variety of *Boolean De Morgan algebras* (viz, De Morgan algebras being also Boolean algebras; cf. [28], [33]) and that of *Boolean bilattices with negation* (viz., expansions of Boolean De Morgan algebras by bilattice connectives; cf. [28]) are equally λ_{DML}-implicative.

3.2.1 De Morgan algebras

A *De Morgan bounded lattice* (viz., *De Morgan algebra*; cf. [2]) is any $\widetilde{F}_{01}$-algebra, whose $\widetilde{F}$-reduct is a De Morgan lattice, while F_{01}-reduct is a bounded distributive lattice, the variety of all them being denoted by DMA. Note that the following identities are true in De Morgan algebras:

$$(3.32) \qquad \sim\!\perp \approx \top,$$
$$(3.33) \qquad \sim\!\top \approx \perp,$$

By $\mathfrak{DMB}_4$ we denote the expansion of both $\mathfrak{DM}_4$ and $(\mathfrak{BD}_2)^2$ to $\widetilde{F}_{01}$.

Lemma 3.24. *Let $\mathfrak{B}$ be an $\widetilde{F}_{01}$-algebra and X a $\wedge$-filter of $\mathfrak{B}$ being both $\sim$-involutive and $\vee$-prime in it and containing $\top^{\mathfrak{B}}$ but not containing $\perp^{\mathfrak{B}}$. Suppose $(\sim^{\mathfrak{B}})^{-1}[X]$ is a $\vee$-filter of $\mathfrak{B}$ being $\wedge$-prime in it and containing $\perp^{\mathfrak{B}}$ but not containing $\top^{\mathfrak{B}}$. Then, $h_{\mathfrak{B}}^{X} \in \operatorname{Hom}(\mathfrak{B}, \mathfrak{DMB}_4)$.*

Proof. In that case, $B \setminus (\sim^{\mathfrak{B}})^{-1}[X]$ is a $\wedge$-filter of $\mathfrak{B}$ being $\vee$-prime in it and containing $\top^{\mathfrak{B}}$ but not containing $\bot^{\mathfrak{B}}$. Hence, by Lemma 3.6, $h_{\mathfrak{B}}^X \in \mathrm{Hom}(\mathfrak{B} \upharpoonright F_{01}, (\mathfrak{B}\mathfrak{D}_2)^2)$. Then, Lemma 3.16 completes the argument. $\qquad\square$

Corollary 3.25 ([2]). $\mathrm{DMA} = \mathbf{I}(\mathfrak{D}\mathfrak{M}\mathfrak{B}_4)$.

Proof. The inclusion from right to left is trivial, for $\mathfrak{D}\mathfrak{M}\mathfrak{B}_4 \in \mathrm{DMA}$. To prove the converse, consider any $\mathfrak{B} \in \mathrm{DMA}$. Take arbitrary distinct $a, b \in B$. Then, $c \triangleq a \wedge^{\mathfrak{B}} b \not\leqslant^{\mathfrak{B}} d \triangleq a \vee^{\mathfrak{B}} b$. Hence, by the Prime Ideal Theorem for distributive lattices, there is a $\wedge$-filter X of $\mathfrak{B}$ $\vee$-prime in it such that $c \in X \not\ni d$, in which case $\top^{\mathfrak{B}} \in X \not\ni \bot^{\mathfrak{B}}$, by (3.6), (3.7). Moreover, by (3.19), X is $\sim$-involutive in $\mathfrak{B}$. And what is more, by (3.20) and (3.21), $(\sim^{\mathfrak{B}})^{-1}[X]$ is a $\vee$-filter of $\mathfrak{B}$ being $\wedge$-prime in it, while, by (3.32) and (3.33), $\bot^{\mathfrak{B}} \in ((\sim^{\mathfrak{B}})^{-1}[X]) \not\ni \top^{\mathfrak{B}}$. Then, by Lemma 3.24, $h_{\mathfrak{B}}^X \in \mathrm{Hom}(\mathfrak{B}, \mathfrak{D}\mathfrak{M}\mathfrak{B}_4)$, while $\pi_0 h_{\mathfrak{B}}^X(c) = 1 \neq 0 = \pi_0 h_{\mathfrak{B}}^X(d)$, in which case $h_{\mathfrak{B}}^X(a) \neq h_{\mathfrak{B}}^X(b)$, as required. $\qquad\square$

The $\widetilde{F}_{01}$-matrix $\mathcal{A} \triangleq \langle \mathfrak{D}\mathfrak{M}\mathfrak{B}_4, (\pi_0)^{-1}[\{1\}] \rangle$ with the same equality determinant $\widetilde{\Upsilon}$ (see above) defines Belnap's four-valued logic with truth constants (cf. [28]). Clearly, for each $i \in 2$, $\gamma_i^{\mathcal{A}} = 0$. The $\widetilde{F}_{01}$-sequent calculus $\mathbb{G}$ of type $[\omega, \omega]$, associated with $\mathcal{A}$ according to [38], can be chosen to consist exactly of classical, De Morgan, involution and structural rules of the same type together with the axioms (2.31), (2.32), (3.8) and (3.9). It was introduced and studied in [28] and [43]. According to the cited works, $\mathbb{G}$ is not algabraizable (cf. [39]).

Though the following result is a particular case of (A.2), to demonstrate applicability and usefulness of Theorem 2.23, we give a direct (and quite transparent) proof of the result under consideration based upon the mentioned theorem.

Proposition 3.26. $\vdash_{\mathbb{G}} = \vdash_{\mathcal{A}^{[\omega,\omega]}}$.

Proof. First, by Theorem 2.23(ii)$\Rightarrow$(i), structural rules are true in $\mathfrak{A}^{[\omega,\omega]}$. Further, remark that $D^{\mathcal{A}}$ is a $\wedge$-filter of $\mathfrak{A}$ being both $\vee$-prime and $\sim$-involutive in it and containing $\top^{\mathfrak{A}}$ but not containing $\bot^{\mathfrak{A}}$, while $(\sim^{\mathfrak{A}})^{-1}[D^{\mathcal{A}}]$ is a $\vee$-filter of $\mathfrak{A}$ being $\wedge$-prime in it and containing $\bot^{\mathfrak{A}}$ but not containing $\top^{\mathfrak{A}}$. Therefore, the axioms (2.19)—(2.30), (2.17), (2.18), (3.8), (3.9), (2.31) and (2.32) are true in $\mathcal{A}^{[\omega,\omega]}$. In particular, in view of Remarks 2.28 and 2.31, classical, De Morgan and involution rules are true in $\mathfrak{A}^{[\omega,\omega]}$. Thus, $\vdash_{\mathbb{G}} \subseteq \vdash_{\mathcal{A}^{[\omega,\omega]}}$. Conversely, by Theorem 2.23(i)$\Rightarrow$(ii), there is some class K of $\widetilde{F}_{01}$-matrices such that $\vdash_{\mathbb{G}} = \vdash_{\mathsf{K}^{[\omega,\omega]}}$. Consider any $\mathcal{B} \in \mathsf{K}$. Then, in view of Remarks 2.28 and 2.31 , the axioms (2.19)—(2.30), (2.17), (2.18), (3.8), (3.9), (2.31) and (2.32) are true in $\mathcal{B}^{[\omega,\omega]}$. Hence, $D^{\mathcal{B}}$ is a $\wedge$-filter of $\mathfrak{B}$ being both $\sim$-involutive and $\vee$-prime in it and containing $\top^{\mathfrak{B}}$ but not containing $\bot^{\mathfrak{B}}$, while $(\sim^{\mathfrak{B}})^{-1}[D^{\mathcal{B}}]$ is a $\vee$-filter of $\mathfrak{A}$ being

49

$\wedge$-prime in it and containing $\perp^{\mathcal{B}}$ but not containing $\top^{\mathcal{B}}$. Therefore, in view of Lemma 3.24, we have $h_{\mathcal{B}} \in \mathrm{SHom}(\mathcal{B}, \mathcal{A})$, and so $h_{\mathcal{B}} \in \mathrm{SHom}(\mathcal{B}^{[\omega,\omega]}, \mathcal{A}^{[\omega,\omega]})$, in view of Remark 2.22. Then, by Lemma 1.14 of [42], $\vdash_{\mathcal{A}^{[\omega,\omega]}} \subseteq \vdash_{\mathcal{B}^{[\omega,\omega]}}$. Thus, $\vdash_{\mathcal{A}^{[\omega,\omega]}} \subseteq \vdash_{\mathbb{G}}$, as required. $\qquad\square$

Lemma 3.27. $\mathbb{G}$ *is weakly contraposable.*

Proof. First, (2.11) immediately follows from (2.15) and (2.16). The rest of argumentation is straightforward with using (2.15), (2.16) and Remark 2.28. $\qquad\square$

Clearly, $\mathbb{G}$ is normal. Then, by Corollary 2.33 and Lemma 3.27, we immediately obtain:

Theorem 3.28. $\widehat{\mathbb{G}}$ *has DT with respect to* $\widetilde{\lambda_{F_{01}^{[\omega,\omega]}}}$.

The following result was first proved in [28] (cf. [43]) with using the proof-theoretical characterization of equivalence provided by Proposition 2.4 of [42] (extending an appropriate result of [27]). On the other hand, it equally follows from Lemma 2.10, Theorem 3.15 and Proposition 3.24 of [44] together with Proposition 3.26, Corollary 3.25 and the fact that $\{1\}$ is a $\wedge$-filter of $\mathfrak{D}_2$ $\vee$-prime in it (this provides a much more transparent algebraic argumentation of the result under consideration).

Proposition 3.29. *Both* $\mathfrak{DMB}_4$ *and* DMA *are finitary relationally equivalent purely-algebraic semantics for* $\widehat{\mathbb{G}}$ *with respect to* $\varepsilon_{\{x_0\}}$ *and* δ_{01}.

3.2.2   Boolean De Morgan algebras

We will also use the following secondary binary connectives (cf. Section 3.3 of [44]):

$$x_0 \equiv x_1 \triangleq (x_0 \leftrightarrows x_1) \wedge (\sim x_0 \leftrightarrows \sim x_1),$$
$$x_0 \Rightarrow x_1 \triangleq (\neg x_0 \vee \sim x_0 \vee (x_1 \wedge \neg \sim x_1)).$$

A *Boolean De Morgan algebra* (cf. [28], [33]) is any $\widetilde{F}_{\mathrm{B}}$-algebra, whose $\widetilde{F}$-reduct is a De Morgan lattice, while F_{B}-reduct is a Boolean algebra, the variety of all them being denoted by BDMA. Boolean De Morgan algebras also satisfy the following identity:

(3.34) $$\sim \neg x \approx \neg \sim x.$$

By $\mathfrak{BDM}_4$ we denote the expansion of both $\mathfrak{DM}_4$ and $(\mathfrak{B}_2)^2$ to $\widetilde{F}_{\mathrm{B}}$.

Lemma 3.30. *Let $\mathfrak{B}$ be an $\widetilde{F}_B$-algebra and X a $\wedge$-filter of $\mathfrak{B}$ being both $\sim$-involutive, $\neg$-complementary and $\vee$-prime in it and containing $\top^{\mathfrak{B}}$ but not containing $\bot^{\mathfrak{B}}$. Suppose $(\sim^{\mathfrak{B}})^{-1}[X]$ is a $\vee$-filter of $\mathfrak{B}$ being both $\neg$-complementary and $\wedge$-prime in it and containing $\bot^{\mathfrak{B}}$ but not containing $\top^{\mathfrak{B}}$. Then, $h_{\mathfrak{B}}^{X} \in \mathrm{Hom}(\mathfrak{B}, \mathfrak{B}\mathfrak{D}\mathfrak{M}_4)$.*

Proof. In that case, $B \setminus (\sim^{\mathfrak{B}})^{-1}[X]$ is a $\wedge$-filter of $\mathfrak{B}$ being both $\neg$-complementary and $\vee$-prime in it and containing $\top^{\mathfrak{B}}$ but not containing $\bot^{\mathfrak{B}}$. Hence, by Lemma 3.10, $h_{\mathfrak{B}}^{X} \in \mathrm{Hom}(\mathfrak{B} \upharpoonright F_B, (\mathfrak{B}_2)^2)$. Then, Lemma 3.16 completes the argument. $\square$

Corollary 3.31 ([28], [33]). BDMA $= \mathbf{I}(\mathfrak{B}\mathfrak{D}\mathfrak{M}_4)$.

Proof. The inclusion from right to left is trivial, for $\mathfrak{B}\mathfrak{D}\mathfrak{M}_4 \in$ BDMA. To prove the converse, consider any $\mathfrak{B} \in$ BDMA. Take arbitrary distinct $a, b \in B$. Then, $c \triangleq a \wedge^{\mathfrak{B}} b \not\leq^{\mathfrak{B}} d \triangleq a \vee^{\mathfrak{B}} b$. Hence, by the Prime Ideal Theorem for distributive lattices, there is a $\wedge$-filter X of $\mathfrak{B}$ $\vee$-prime in it such that $c \in X \not\ni d$, in which case $\top^{\mathfrak{B}} \in X \not\ni \bot^{\mathfrak{B}}$, by (3.6), (3.7), while, by (3.11) and (3.12), X is $\neg$-complementary in $\mathfrak{B}$. Moreover, by (3.19), X is $\sim$-involutive in $\mathfrak{B}$. And what is more, by (3.20) and (3.21), $(\sim^{\mathfrak{B}})^{-1}[X]$ is a $\vee$-filter of $\mathfrak{B}$ being $\wedge$-prime in it, while, by (3.32) and (3.33), $\bot^{\mathfrak{B}} \in ((\sim^{\mathfrak{B}})^{-1}[X]) \not\ni \top^{\mathfrak{B}}$, whereas, by (3.34), $(\sim^{\mathfrak{B}})^{-1}[X]$ is $\neg$-complementary. Then, by Lemma 3.30, $h_{\mathfrak{B}}^{X} \in \mathrm{Hom}(\mathfrak{B}, \mathfrak{B}\mathfrak{D}\mathfrak{M}_4)$, while $\pi_0 h_{\mathfrak{B}}^{X}(c) = 1 \neq 0 = \pi_0 h_{\mathfrak{B}}^{X}(d)$, in which case $h_{\mathfrak{B}}^{X}(a) \neq h_{\mathfrak{B}}^{X}(b)$, as required. $\square$

The $\widetilde{F}_B$-matrix $\mathcal{A} \triangleq \langle \mathfrak{B}\mathfrak{D}\mathfrak{M}_4, (\pi_0)^{-1}[\{1\}] \rangle$ with the same equality determinant $\widetilde{\Upsilon}$ (see above) defines Belnap's four-valued logic with classical negation $\mathbb{BB}_4$ (cf. [28], [43]). Clearly, for each $i \in 2$, $\gamma_i^{\mathcal{A}} = 0$. The $\widetilde{F}_B$-sequent calculus $\mathbb{G}$ of type $[\omega, \omega]$, associated with $\mathcal{A}$ according to [38], can be chosen to consist exactly of classical, De Morgan, structural rules and rules (2.15) and (2.16) of the same type together with the axioms (2.31), (2.32), (3.8), (3.9) as well as the rules (3.13) and (3.14) collectively with the following additional rules (where $m, n \in \omega$):

$$(3.35) \qquad \frac{\bar{x}(1, m) \rightarrowtail \bar{x}(m+1, n), \sim x_0}{\sim \neg x_0, \bar{x}(1, m) \rightarrowtail \bar{x}(m+1, n)},$$

$$(3.36) \qquad \frac{\sim x_0, \bar{x}(m+1, n) \rightarrowtail \bar{x}(1, m)}{\bar{x}(m+1, n) \rightarrowtail \sim \neg x_0, \bar{x}(1, m)}.$$

51

It was introduced and studied in [28] and [43]. (Recall that structural and classical rules together with both (3.13) and (3.14) constitute the propositional fragment of Gentzen's LK [9].) According to the cited works, $\mathbb{G}$ is not algabraizable (cf. [39]).

Remark 3.32. Under presence of structural rules, (3.35) and (3.36) are reversible. More precisely, the following axioms are immediately derivable from Reflexivity by means of (3.35) and (3.36) of type $[\omega, \omega]$, respectively:

$$(3.37) \qquad\qquad {\sim}\neg x_0, {\sim}x_0 \quad \longmapsto \quad ,$$

$$(3.38) \qquad\qquad\qquad\qquad \longmapsto \quad {\sim}\neg x_0, {\sim}x_0.$$

On the other hand, the rules inverse to (3.35) and (3.36) of the same type (as well as the latter themselves) are derivable from the axioms (3.37) and (3.38) by means of structural rules of the same type. Moreover, these axioms are equally derivable from Reflexivity by means of the rules inverse to (3.35) and (3.36) of the same type. In particular, the rules (3.35) and (3.36) of the same type and those which are inverse to them are, so to say, inter-derivable modulo structural rules of the same type. □

Though the following result is a particular case of (A.2), to demonstrate applicability and usefulness of Theorem 2.23, we give a direct (and quite transparent) proof of the result under consideration based upon the mentioned theorem.

Proposition 3.33. $\vdash_{\mathbb{G}} = \vdash_{\mathcal{A}^{[\omega,\omega]}}$.

Proof. First, by Theorem 2.23(ii)$\Rightarrow$(i), structural rules are true in $\mathfrak{A}^{[\omega,\omega]}$. Further, remark that $D^{\mathcal{A}}$ is a $\wedge$-filter of $\mathfrak{A}$ being both $\vee$-prime, $\neg$-complementary and ${\sim}$-involutive in it and containing $\top^{\mathfrak{A}}$ but not containing $\perp^{\mathfrak{A}}$, whereas

$$({\sim}^{\mathfrak{A}})^{-1}[D^{\mathcal{A}}]$$

is a $\vee$-filter of $\mathfrak{A}$ being both $\neg$-complementary and $\wedge$-prime in it and containing $\perp^{\mathfrak{A}}$ but not containing $\top^{\mathfrak{A}}$. Therefore, the axioms (2.19)—(2.30), (2.17), (2.18), (3.15), (3.16), (3.37), (3.38), (3.8), (3.9), (2.31) and (2.32) are true in $\mathcal{A}^{[\omega,\omega]}$. In particular, in view of Remarks 2.28, 2.31, 3.12 and 3.32, classical, De Morgan and involution rules as well as the rules (3.13), (3.14), (3.35), (3.36) are true in $\mathfrak{A}^{[\omega,\omega]}$. Thus, $\vdash_{\mathbb{G}} \subseteq \vdash_{\mathcal{A}^{[\omega,\omega]}}$. Conversely, by Theorem 2.23(i)$\Rightarrow$(ii), there is some class K of $\widetilde{F}_{\mathrm{B}}$-matrices such that $\vdash_{\mathbb{G}} = \vdash_{\mathrm{K}^{[\omega,\omega]}}$. Consider any $\mathcal{B} \in$ K. Then, in view of Remarks 2.28, 2.31, 3.12 and 3.32, the axioms (2.19)—(2.30), (2.17), (2.18), (3.15), (3.16), (3.37), (3.38), (3.8), (3.9), (2.31) and (2.32) are true in $\mathcal{B}^{[\omega,\omega]}$. Hence, $D^{\mathcal{B}}$ is a $\wedge$-filter of $\mathfrak{B}$ being both ${\sim}$-involutive, $\neg$-complementary and $\vee$-prime in it and containing $\top^{\mathcal{B}}$ but not containing $\bot^{\mathcal{B}}$, while $(\sim^{\mathcal{B}})^{-1}[D^{\mathcal{B}}]$ is a $\vee$-filter of $\mathfrak{A}$ being both $\neg$-complementary and $\wedge$-prime in it and containing $\bot^{\mathcal{B}}$ but not containing $\top^{\mathcal{B}}$. Therefore, in view of Lemma 3.30, we have $h_{\mathcal{B}} \in \mathrm{SHom}(\mathcal{B}, \mathcal{A})$, and so $h_{\mathcal{B}} \in \mathrm{SHom}(\mathcal{B}^{[\omega,\omega]}, \mathcal{A}^{[\omega,\omega]})$, in view of Remark 2.22. Then, by Lemma 1.14 of [42], $\vdash_{\mathcal{A}^{[\omega,\omega]}} \subseteq \vdash_{\mathcal{B}^{[\omega,\omega]}}$. Thus, $\vdash_{\mathcal{A}^{[\omega,\omega]}} \subseteq \vdash_{\mathbb{G}}$, as required. $\qquad\square$

Lemma 3.34. $\mathbb{G}$ *is weakly contraposable.*

Proof. First, (2.11) immediately follows from (2.15) and (2.16). The rest of argumentation is straightforward with using (2.15), (2.16) and Remark 2.28. $\qquad\square$

Clearly, $\mathbb{G}$ is normal. Then, by Corollary 2.33 and Lemma 3.34, we immediately obtain:

Theorem 3.35. $\widehat{\mathbb{G}}$ *has DT and holds PL with respect to* $\lambda_{\widetilde{F}_{\mathrm{B}}^{[\omega,\omega]}}$.

Since $D^{\mathcal{A}} \not\ni \bot^{\mathfrak{A}}$ is both $\vee$-prime and $\neg$-complementary in $\mathfrak{A}$, it is easy to see that:

$$
(3.39) \qquad \mathcal{A}^{[\omega,\omega]} \;=\; \vartheta^{-1}[\mathcal{A}],
$$
$$
(3.40) \qquad \mathcal{A} \;=\; \varrho^{-1}[\mathcal{A}^{[\omega,\omega]}].
$$

By Theorem 2.10(v)$\Rightarrow$(i) of [42] together with Proposition 3.33, (3.39) and (3.40), we immediately get:

Proposition 3.36 (cf. [43]). $\mathbb{BB}_4$ *and* $\mathbb{G}$ *are relationally equivalent with respect to* ϱ *and* ϑ.

According to (3.35), (3.39), (3.40) and Corollary 3.61 of [44], the logic $\mathbb{TSBB}_4$ of the $\widetilde{F}_{\mathrm{B}}$-matrix $\mathcal{B} \triangleq \langle \mathfrak{BDM}_4, \{\langle 1,1 \rangle\} \rangle$ is the extension of $\mathbb{BB}_4$ relatively axiomatized by the single rule:

$$
(3.41) \qquad D(x) \to D(x \equiv \top).
$$

Corollary 3.37. $\mathbb{TSBB}_4$ *and* $\widehat{\mathbb{G}}$ *are relationally equivalent with respect to* ϱ *and* ϑ.

Proof. Recall that the following universal sentence:

$$
(3.42) \qquad (x \lessgtr y) \leftrightarrow ((x \supset y) \approx \top)
$$

is true in Boolean algebras, while the quasi-identity:

$$(x \lesssim y) \rightarrow (\sim y \lesssim \sim x)$$

is true in De Morgan lattices. Hence, the rule $\vartheta(2.10)$ is true in $\mathcal{B}$. Moreover, it is quite straightforward to verify that the rule $\varrho(3.41)$ is derivable in $\widehat{\mathbb{G}}$. Then, Proposition 3.36 and Corollary 1.1 complete the argument. $\qquad \square$

Corollary 3.38. $\mathbb{TSBB}_4$ *has DT and holds PL with respect to* $\{D(x_0 \Rrightarrow x_1)\}$.

Proof. Applying Theorem 2.8 together with (2.14), by Corollary 3.37 and Theorem 3.35, we immediately conclude that $\mathbb{TSBB}_4$ has DT and holds PL with respect to:

$$\vartheta[\widetilde{\lambda_{F_{\mathrm{B}}^{[\omega,\omega]}}}(\varrho(D), \varrho(D)[x_0/x_1])]$$

$$= \vartheta[\widetilde{\lambda_{F_{\mathrm{B}}^{[\omega,\omega]}}}((\rightarrowtail x_0), (\rightarrowtail x_1))]$$

$$= \vartheta[\{x_0 \rightarrowtail x_1, \sim x_0; x_0, \sim x_1 \rightarrowtail \sim x_0\}]$$

$$\dashv\vdash_{\mathcal{B}} \{D(x_0 \Rrightarrow x_1)\}.$$

Then, Proposition 2.1 and Corollary 2.5 complete the argument. $\qquad \square$

Remark 3.39. It is easy to see that $\mathcal{B}$ is both $\{x_0 \Rrightarrow x_1\}$-prime and -implicative. In view of Theorem 2.4(iv)$\Rightarrow$(i), this immediately implies Corollary 3.38. However, such am immediate argumentation (as opposed to the constructive one presented above), would hide genesis of such a quite extraordinary and unexpected secondary implication as $\Rrightarrow$. $\qquad \square$

Consider the purely relational parameter-less $(\langle F_{\mathrm{B}}, \{D\}\rangle, (F_{\mathrm{B}} + \approx))$-translation τ, given by $\tau(D) \triangleq \{x_0 \approx \top\}$, and $((F_{\mathrm{B}} + \approx), \langle F_{\mathrm{B}}, \{D\}\rangle)$-translation ρ, defined by $\rho(\approx) \triangleq \{D(x_0 \leftrightharpoons x_1)\}$. Obviously, we have:

(3.43) $$\mathcal{B} = \tau^{-1}[\mathfrak{B} + \approx].$$

Moreover, by (3.42), we immediately get:

(3.44) $$(\mathfrak{B} + \approx) = \rho^{-1}[\mathcal{B}].$$

By Theorem 2.10(v)$\Rightarrow$(i) of [42], Corollary 3.31, (3.43) and (3.44) immediately yield:

Proposition 3.40 (cf. Remark 3.60 of [44]). *Both* $\mathfrak{BDM}_4$ *and* BDMA *are finitary relationally equivalent purely-algebraic semantics for* $\mathbb{TSBB}_4$ *with respect to* ρ *and* τ.

The following result was first proved in [28] (cf. [43]) with using the proof-theoretical characterization of equivalence provided by Proposition 2.4 of [42] (extending an appropriate result of [27]). Below, we present a much more transparent and easier argumentation of the result under consideration based upon the above results and the transitivity of equivalence between UHTs.

Corollary 3.41. *Both $\mathfrak{BDM}_4$ and BDMA are finitary relationally equivalent purely-algebraic semantics for $\widehat{\mathbb{G}}$ with respect to $\varepsilon_{\{x_0\}}$ and δ_{01}.*

Proof. In view of Proposition 2.9 of [42], by Corollary 3.37 and Proposition 3.40, we first conclude that both $\mathfrak{BD}_4$ and BDMA are finitary relationally equivalent purely-algebraic semantics for $\widehat{\mathbb{G}}$ with respect to $\varrho \circ \rho$ and $\tau \circ \vartheta$.[2] On the other hand, by (3.42), we have $\delta_{01} \cong_{\text{BDMA}} \tau \circ \vartheta$. Moreover, with using Remarks 2.31 and 3.12, it is routine checking that $\varepsilon_{\{x_0\}} \cong_{\widehat{\mathbb{G}}} \varrho \circ \rho$. Then, Proposition 1.3 completes the argument. $\qquad\square$

3.3   Stone algebras

Recall that a *Stone algebra* (cf. [2], [11], [12]) is any $\widetilde{F}_{01}$-algebra, whose F_{01}-reduct is a bounded distributive lattice, while the identity (3.32) together with the following ones are true in it:[3]

$$(3.45) \qquad\qquad x \wedge {\sim}x \approx \perp,$$

$$(3.46) \qquad\qquad x \wedge {\sim}(x \wedge y) \approx x \wedge {\sim}y,$$

$$(3.47) \qquad\qquad {\sim}x \vee {\sim}{\sim}x \approx \top,$$

the variety of all them being denoted by SA. Recall that the *De Morgan* identities (3.20) and (3.21) are true in Stone algebras.[4]

Given any $n \in \omega$, let $\mathfrak{S}_n$ be the expansion of $\mathfrak{BD}_n$ to $\widetilde{F}_{01}$ given by

$$\sim^{\mathfrak{S}_n} a \triangleq \max\{b \in n \mid \min(a, b) = 0\},$$

---

[2]As a composition of purely relational translations is unique, we use standard notation for composition in such cases.

[3]The first two identities together with both (3.32) and the identities axiomatizing the variety of bounded distributive lattices axiomatize the variety of *pseudo-complemented bounded distributive lattices* (cf. [2]), the identity (3.33) being a consequence of (3.45), while (3.47) being a relative axiomatization of the subvariety of Stone algebras (cf. [12]).

[4]The former identity is true in the variety of all pseudo-complemented bounded distributive lattices, the latter being a one more relative axiomatization of the subvariety of Stone algebras (cf. [12]).

for all $a \in n$. Note that:

$$(3.48) \qquad \qquad \mathfrak{S}_2 = \mathfrak{B}_2[\neg/\sim].$$

Consider the mapping $e_3 : 3 \to 2^2, a \mapsto \langle [a/2], \min(1, a) \rangle$. Once e_3 is injective, we have $e_3[\mathfrak{S}_3]$ such that:

$$(3.49) \qquad \mathfrak{B}\mathfrak{D}_3 = (e_3)^{-1}[(\mathfrak{B}\mathfrak{D}_2)^2],$$
$$(3.50) \qquad \sim^{e_3[\mathfrak{S}_3]} \langle a, b \rangle = \langle 1 - b, 1 - b \rangle,$$

for all $\langle a, b \rangle \in \operatorname{img} e_3$.

Lemma 3.42. *Let $\mathfrak{B}$ be an $\widetilde{F}_{01}$-algebra and X a $\wedge$-filter of $\mathfrak{B}$ being $\vee$-prime in it and containing $\top^{\mathfrak{B}}$ but not containing $\bot^{\mathfrak{B}}$. Suppose $(\sim^{\mathfrak{B}})^{-1}[X]$ is a $\vee$-filter of $\mathfrak{B}$ being both $\sim$-complementary and $\wedge$-prime in it, being also disjoint with X and containing $\bot^{\mathfrak{B}}$ but not containing $\top^{\mathfrak{B}}$. Then, $h_{\mathfrak{B}}^X \in \operatorname{Hom}(\mathfrak{B}, e_3[\mathfrak{S}_3])$.*

Proof. In that case, $B \setminus (\sim^{\mathfrak{B}})^{-1}[X]$ is a $\wedge$-filter of $\mathfrak{B}$ being both $\sim$-complementary and $\vee$-prime in it and containing $\top^{\mathfrak{B}}$ but not containing $\bot^{\mathfrak{B}}$. Hence, by Lemma 3.10 and (3.48), $\pi_1 \circ h_{\mathfrak{B}}^X \in \operatorname{Hom}(\mathfrak{B}, \mathfrak{S}_2)$, while, by Lemma 3.6, $\pi_0 \circ h_{\mathfrak{B}}^X \in \operatorname{Hom}(\mathfrak{B} \upharpoonright F_{01}, \mathfrak{B}\mathfrak{D}_2)$. Moreover, as $X \cap (\sim^{\mathfrak{B}})^{-1}[X] = \varnothing$, we have $\langle 1, 0 \rangle \notin h_{\mathfrak{B}}^X[B]$, so $(\operatorname{img} h_{\mathfrak{B}}^X) \subseteq (\operatorname{img} e_3)$. Then, (3.49) and (3.50) complete the argument. $\qquad \square$

Corollary 3.43 (cf. [12]). $\mathbf{SA} = \mathbf{I}(\mathfrak{S}_3)$.

Proof. The inclusion from right to left is trivial, for $\mathfrak{S}_3 \in \mathbf{SA}$. To prove the converse, consider any $\mathfrak{B} \in \mathbf{BDMA}$. Take arbitrary distinct $a, b \in B$. Then, $c \triangleq a \wedge^{\mathfrak{B}} b \not\leqslant^{\mathfrak{B}} d \triangleq a \vee^{\mathfrak{B}} b$. Hence, by the Prime Ideal Theorem for distributive lattices, there is a $\wedge$-filter X of $\mathfrak{B}$ $\vee$-prime in it such that $c \in X \not\ni d$, in which case $\top^{\mathfrak{B}} \in X \not\ni \bot^{\mathfrak{B}}$, by (3.6), (3.7). Moreover, by (3.20), (3.21), (3.45) and (3.47), $(\sim^{\mathfrak{B}})^{-1}[X]$ is a $\vee$-filter of $\mathfrak{B}$ being both $\sim$-complementary and $\wedge$-prime in it being also disjoint with X, while, by (3.32) and (3.33), $\bot^{\mathfrak{B}} \in ((\sim^{\mathfrak{B}})^{-1}[X]) \not\ni \top^{\mathfrak{B}}$. Then, by Lemma 3.42, $(e_3)^{-1} \circ h_{\mathfrak{B}}^X \in \operatorname{Hom}(\mathfrak{B}, \mathfrak{S}_3)$, while $\pi_0 h_{\mathfrak{B}}^X(c) = 1 \neq 0 = \pi_0 h_{\mathfrak{B}}^X(d)$, in which case $(e_3)^{-1} h_{\mathfrak{B}}^X(a) \neq (e_3)^{-1} h_{\mathfrak{B}}^X(b)$, as required. $\qquad \square$

The $\widetilde{F}_{01}$-matrix $\mathcal{A} = \langle \mathfrak{S}_3, 3 \setminus 2 \rangle$, with the same equality determinant $\widetilde{\Upsilon}$ (cf. Section 3.2) defines the implication-less fragment of Dummett's *linear intuitionistic logic* [6]. Clearly, for each $i \in 2$, $\gamma_i^{\mathcal{A}} = 0$. The $\widetilde{F}_{01}$-sequent calculus $\mathbb{G}$ of type $[\omega, \omega]$, associated with $\mathcal{A}$ according to [38], can be chosen to consist exactly of classical, De Morgan, and structural rules of the same type together with the axioms (2.31), (2.32), (3.8), (3.9) as well as the following additional rules and axiom (where $m, n \in \omega$):

(3.51)
$$x_0, {\sim} x_0 \longmapsto ,$$

(3.52)
$$\frac{\bar{x}(m+1, n), {\sim} x_0 \longmapsto \bar{x}(1, m)}{\bar{x}(m+1, n) \longmapsto {\sim}{\sim} x_0, \bar{x}(1, m)},$$

It was actually suggested in [24] and then studied in [27] and [43]. According to the latter two works, $\mathbb{G}$ is not algabraizable (cf. [39]).

Remark 3.44. Under presence of structural rules and the axiom (3.51), the rules (3.52) are reversible. More precisely, the following axiom is immediately derivable from Reflexivity by means of (3.52) of type $[\omega, \omega]$:

(3.53)
$$\longmapsto \quad {\sim}{\sim} x_0, {\sim} x_0.$$

On the other hand, the rules inverse to (3.52) of the same type (resp., the latter themselves) are derivable from the axiom (3.51) (resp., (3.53)) by means of structural rules of the same type. □

Though the following result is a particular case of (A.2), to demonstrate applicability and usefulness of Theorem 2.23, we give a direct (and quite transparent) proof of the result under consideration based upon the mentioned theorem.

Proposition 3.45. $\vdash_{\mathbb{G}} = \vdash_{A^{[\omega, \omega]}}$.

Proof. First, by Theorem 2.23(ii)⇒(i), structural rules are true in $\mathfrak{A}^{[\omega, \omega]}$. Further, remark that D^A is a $\wedge$-filter of $\mathfrak{A}$ being $\vee$-prime in it and containing $\top^{\mathfrak{A}}$ but not containing $\bot^{\mathfrak{A}}$, while $({\sim}^{\mathfrak{A}})^{-1}[D^A]$ is a $\vee$-filter of $\mathfrak{A}$ being both ${\sim}$-complementary and $\wedge$-prime in it as well as disjoint with D^A and containing $\bot^{\mathfrak{A}}$ but not containing $\top^{\mathfrak{A}}$. Therefore, the axioms (2.19)—(2.30), (3.51), (3.53), (3.8), (3.9), (2.31) and (2.32) are true in $A^{[\omega, \omega]}$. In particular, in view of Remarks 2.31 and 3.44, classical and De Morgan rules as well as the rules (3.52) are true in $\mathfrak{A}^{[\omega, \omega]}$. Thus, $\vdash_{\mathbb{G}} \subseteq \vdash_{A^{[\omega, \omega]}}$. Conversely, by Theorem 2.23(i)⇒(ii), there is some class K of $\widetilde{F_{01}}$-matrices such that $\vdash_{\mathbb{G}} = \vdash_{K^{[\omega, \omega]}}$. Consider any $\mathcal{B} \in$ K. Then, in view of Remarks 2.31 and 3.44, the axioms (2.19)—(2.30), (3.51), (3.53), (3.8), (3.9), (2.31) and (2.32) are true in $\mathcal{B}^{[\omega, \omega]}$. Hence, D^B is a $\wedge$-filter of $\mathfrak{B}$ being $\vee$-prime in it and containing $\top^{\mathfrak{B}}$ but not containing $\bot^{\mathfrak{B}}$, while $({\sim}^{\mathfrak{B}})^{-1}[D^B]$ is a $\vee$-filter of $\mathfrak{A}$ being both ${\sim}$-complementary and $\wedge$-prime in it as well as disjoint with D^B and containing $\bot^{\mathfrak{B}}$ but not containing $\top^{\mathfrak{B}}$.

Therefore, in view of Lemma 3.42, we have $(e_3)^{-1} \circ h_\mathcal{B} \in \mathrm{SHom}(\mathcal{B}, \mathcal{A})$, and so $(e_3)^{-1} \circ h_\mathcal{B} \in \mathrm{SHom}(\mathcal{B}^{[\omega,\omega]}, \mathcal{A}^{[\omega,\omega]})$, in view of Remark 2.22. Then, by Lemma 1.14 of [42], $\vdash_{\mathcal{A}^{[\omega,\omega]}} \subseteq \vdash_{\mathcal{B}^{[\omega,\omega]}}$. Thus, $\vdash_{\mathcal{A}^{[\omega,\omega]}} \subseteq \vdash_\mathbb{G}$, as required. $\qquad\square$

Lemma 3.46. $\mathbb{G}$ *is weakly contraposable.*

Proof. First, (2.11) immediately follows from (3.51), (3.52) and structural rules. The rest of argumentation is straightforward with using (3.51), (3.53), structural rules and Remark 3.44. $\qquad\square$

Clearly, $\mathbb{G}$ is normal. Then, by Corollary 2.33 and Lemma 3.46, we immediately obtain:

Theorem 3.47. $\widehat{\mathbb{G}}$ *has DT with respect to* $\lambda_{\widetilde{F}_{01}^{[\omega,\omega]}}$.

The following result was first proved in [27] (cf. [43]) with using the proof-theoretical characterization of equivalence provided by Proposition 2.4 of [42] (extending an appropriate result of [27] itself). On the other hand, it equally follows from Lemma 2.10, Theorem 3.4 and the argumentation of Proposition 3.24 (equally applicable to PSA instead of PDML) of [44] together with Proposition 3.45, Corollary 3.43 and the fact that $3 \setminus 2$ is a $\wedge$-filter of $\mathfrak{S}_3$ being $\vee$-prime in it (this provides a much more transparent algebraic argumentation of the result under consideration).

Proposition 3.48. *Both* $\mathfrak{S}_3$ *and* SA *are finitary relationally equivalent purely-algebraic semantics for* $\widehat{\mathbb{G}}$ *with respect to* $\varepsilon_{\{x_0\}}$ *and* δ_{01}.

Consider the following F-equations over V^4:

$$(3.54) \qquad \sim x_0 \wedge \sim x_1 \wedge x_2 \lessgtr x_3,$$
$$(3.55) \qquad \sim x_0 \wedge \sim x_1 \wedge x_3 \lessgtr x_2.$$

Put $\lambda_{\mathsf{SA}} \triangleq \{(3.22), (3.54), (3.3), (3.26), (3.55), (3.5)\}$.

Lemma 3.49. $\lambda_{\mathsf{SA}} \vdash_{\mathsf{SA}}^{\approx} \sigma[\lambda_{\mathsf{SA}}]$.

Proof. First, in view of (3.45), the following hold:

$$(3.56) \qquad ((x \wedge y) \lessgtr z) \vdash_{\mathsf{SA}}^{\approx} ((x \wedge \sim z) \lessgtr \sim y).$$

Hence, the following hold:

(3.57)	(3.54) $\vdash_{\mathsf{SA}}^{\approx}$ $\sigma(3.55)$,
(3.58)	(3.3) $\vdash_{\mathsf{SA}}^{\approx}$ $\sigma(3.5)$,
(3.59)	(3.55) $\vdash_{\mathsf{SA}}^{\approx}$ $\sigma(3.54)$,
(3.60)	(3.5) $\vdash_{\mathsf{SA}}^{\approx}$ $\sigma(3.3)$.

Finally, since the mapping $\delta \restriction \mathrm{Seq}_{\sim}^{[\omega\backslash 1,\omega\backslash 1]}(V^4)$ is injective, the following definitions are correct:

$$\widetilde{\tau(\Phi)} \triangleq \tau(\widetilde{\Phi}),$$
$$\widetilde{\tau(\Phi)} \sqsubseteq \widetilde{\tau(\Psi)} \iff \widetilde{\Phi} \sqsubseteq \widetilde{\Psi}.$$

where $\Phi, \Psi \in \mathrm{Seq}_{\sim}^{[\omega\backslash 1,\omega\backslash 1]}(V^4)$. By the De Morgan identities (3.20) and (3.21), the following hold:

(3.61)
$$\Phi \vdash_{\mathrm{SA}}^{\approx} \widetilde{\Phi},$$

(3.62)
$$(\Phi \sqsubseteq \Psi) \implies (\Phi \vdash_{\mathrm{SA}}^{\approx} \Psi),$$

where $\Phi, \Psi \in \tau[\mathrm{Seq}_{\sim}^{[\omega\backslash 1,\omega\backslash 1]}(V^4)]$. Clearly, we have:

$$\widetilde{(3.5)} \sqsubseteq \sigma(3.22),$$
$$\widetilde{(3.3)} \sqsubseteq \sigma(3.26).$$

Then, by (3.61) and (3.62), we get:

(3.63)
$$(3.5) \vdash_{\mathrm{SA}}^{\approx} \sigma(3.22).$$

(3.64)
$$(3.3) \vdash_{\mathrm{SA}}^{\approx} \sigma(3.26).$$

In this way, (3.57)–(3.60) together with (3.63) and (3.64) complete the argument. $\qquad\square$

Corollary 3.50 (cf. [17]). SA *has REDPC with respect to* λ_{SA}.

Proof. Using Theorem 2.8, Proposition 3.48 and Corollary 2.11, together with (2.14) and omitting tautological (in Stone algebras) equations (viz., identities true in SA) of the form $\delta(\overline{\varphi} \rightarrowtail \overline{\psi})$, where $(\overline{\varphi} \rightarrowtail \overline{\psi}) \in \mathrm{Seq}_{\sim}^{[\omega\backslash 1,\omega\backslash 1]}(V^4)$, while $(\mathrm{img}\,\overline{\varphi}) \cap (\mathrm{img}\,\overline{\psi}) \neq \varnothing$, we first conclude that SA has REDPC with respect to $\lambda_{\mathrm{DML}} \cup \sigma[\lambda_{\mathrm{DML}}]$. First, by (3.45) and (3.62), the following hold:

(3.65)
$$(3.23) \dashv\vdash_{\mathrm{SA}}^{\approx} (3.54),$$

(3.66)
$$(3.24) \dashv\vdash_{\mathrm{SA}}^{\approx} (3.3),$$

(3.67)
$$(3.27) \dashv\vdash_{\mathrm{SA}}^{\approx} (3.55),$$

(3.68)
$$(3.28) \dashv\vdash_{\mathrm{SA}}^{\approx} (3.5).$$

Moreover, both (3.25) and (3.29) are identities true in Stone algebras, in view of (3.45). This together with (3.65)–(3.68) imply that SA has REDPC with respect to $\lambda_{\mathrm{SA}} \cup \sigma[\lambda_{\mathrm{SA}}]$. Then, Lemma 3.49 completes the argument. $\qquad\square$

By (3.32), (3.56) and Proposition 3.48, the following rule (where, generally speaking, $m, n \in \omega$) is derivable in $\widehat{\mathbb{G}}$, when $m = 0$:

$$\text{(3.69)} \qquad \frac{\overline{x}(0, n+1) \rightarrowtail \sim\overline{x}(n+1, m)}{\overline{x}(1, n) \rightarrowtail \sim x_0, \sim\overline{x}(n+1, m)}$$

In that case, in view of Remark 3.44, with using (3.53), (3.51) and structural rules, we conclude that (3.69) is derivable in $\widehat{\mathbb{G}}$, for all $m \in \omega$. On the other hand, in view of (3.51) and structural rules, the rule (2.10) is derivable in the extension $\mathbb{LLJ}$ of $\mathbb{G}$ by (3.69), for $m = 0$ alone. Thus, we first have:

$$\text{(3.70)} \qquad \vdash_{\mathbb{LLJ}} = \vdash_{\widehat{G}}.$$

Properly speaking, it was $\mathbb{LLJ}$ that was considered in [27] and [43]. A closely related calculus LLJ^+ with Cut Elimination property has been introduced and studied in [36]. It is constituted by structural, classical rules, the axioms (3.8), (3.9) and the rules (3.13)[$\neg/\sim$] and (3.69), for all $m \in \omega$. It is easy to see that the axioms (2.25)—(2.30) are derivable in LLJ^+, so, in view of Remark 2.31, De Morgan rules are derivable in LLJ^+. Moreover, (3.13)[$\neg/\sim$] and (3.51) are inter-derivable modulo structural rules. Thus, in view of (3.70), we also get:

$$\text{(3.71)} \qquad \vdash_{LLJ^+} = \vdash_{\widehat{G}}.$$

Concluding this section, we prove negative results concerning PL.

A Stone algebra is said to be *Boolean*, if it satisfies the following identity:

$$\text{(3.72)} \qquad x \vee \sim x \approx \top.$$

We start from proving:

Theorem 3.51. SA *is not implicative.*

Proof. By contradiction. For suppose SA is implicative. Then, by Lemma 2.12, Theorem 2.13 and Corollary 3.50, there is some $\mathsf{K} \subseteq \mathsf{SA}$ such that $\mathsf{SA} = \mathbf{I}(\mathsf{K})$, while λ_{SA} is an implication system for K, in which case any algebra embeddable into any member of K is λ_{SA}-implicational, and so algebraically simple. Since $\mathfrak{S}_3 \in \mathsf{SA}$ is not Boolean (for (3.72) is not true in it under the assignment $[x/1]$), K must contain a non-Boolean $\mathfrak{B}$, in which case there must be some $a \in B$ such that $b \triangleq (a \vee^{\mathfrak{B}} \sim^{\mathfrak{B}} a) \neq \top^{\mathfrak{B}}$. Then, by (3.6) and (3.7), $\bot^{\mathfrak{B}} \neq \top^{\mathfrak{B}}$, while, by (3.20) and (3.45), $\sim^{\mathfrak{B}} b = \bot^{\mathfrak{B}}$, and so, by (3.32), $b \neq \bot^{\mathfrak{B}}$. Thus, in view of (3.33), the mapping $e : 3 \to B$, defined by:

$$e2 \triangleq \top^{\mathfrak{B}},$$
$$e1 \triangleq b,$$
$$e0 \triangleq \bot^{\mathfrak{B}},$$

is an embedding of $\mathfrak{S}_3$ into $\mathfrak{B}$. Hence, $\mathfrak{S}_3$ is algebraically simple. However, the mapping $h : 3 \to 2, a \mapsto (1 - [(2 - a)/2])$ is a homomorphism from $\mathfrak{S}_3$ onto $\mathfrak{S}_2$, in which case $\ker h \in \mathrm{Con}(\mathfrak{S}_3) \subseteq \{\Delta_3, 3^2\}$. On the other hand, h is not injective, for $h1 = 1 = h2$, in which case $\ker h \neq \Delta_3$, while $|\operatorname{img} h| = |2| = 2 \neq 1$, and so $\ker h \neq 3^2$. This contradiction completes the argument. $\qquad\square$

As a consequence of Theorem 3.51, Proposition 3.48 and Corollary 2.17, we immediately obtain:

Corollary 3.52. $\widehat{\mathbb{G}}$ *does not hold PL.*

Corollary 3.52 shows that the [optional] condition of $\mathbb{G}$ having involution rules is essential for Corollary 2.30 [2.33] to hold, whereas the condition of the reversibility of Strong Contraposition rules is essential for Theorem 2.27 to hold, while, in view of Theorem 2.19 and Lemmas 2.32 and 3.46, the "Peirce Law strengthening" of Theorem 2.25 is not, generally speaking, valid.

Conclusions

Thus, the main part of the present monograph has successfully solved the tasks raised in the Introduction. And what is more, Section 3.2.2 displays applicability of our generic method to sentential logics as well. On the other hand, there may be more examples to be covered by the generic paradigm underlying this book. And we strongly encourage others to join us for looking for and studying these. It might then be reasonable to do it within the context of the "many-place sequent" extension [40] of both [38] and [39].

It is remarkable that the generic elaboration presented in this book collectively with effective procedures to be extracted from the algebraic study of sequent calculi associated with finitely-valued logics with equality determinant that is due to [39] do yield quite useful computational tools of studying RED-PRC in [implicativity of] respective quasivarieties. This point however require further perfection that is going to be discussed elsewhere.

Concluding our discussion, we should like to highlight also that the many-dimensional formalism, adopted in [29] and underlying orthodox algebraic logic, does not cover multiple-conclusion sequent calculi at all.[5] For this reason, the present study could nowise be elaborated within the Procrustean bad of ortho-

---

[5]In this connection, it is noteworthy that the extension of many-dimensional formalism to the general first-order one is far from being straightforward, as certain quite ambitious "authorities" in orthodox algebraic logic try to claim. The thing is that the former does not involve predicate constants. On the other hand, the empty sequent cannot but be treated as a predicate constant. On the other hand, predicate constants inevitably give rise to appearance of parameter variables in translations that cannot be disregarded or treated just as "dummy" ones (cf. both [27] and [42]). Such parameterization of translations leads to the situation, when substitutions do no longer commute with translations in the absolute sense, but only in a relative one (cf. ibid.). On the other hand, it is such relative commutativity that must be adopted in extending equivalence between UHTs to functional translations (cf. [42]). In this way, the conservatism of orthodox algebraic logic has inevitably prevented from involving functional translations within its context that has become, by now, nothing but Procrustean bad also because of its initial unwillingness to cover prevarieties of algebraic systems but only those of pure algebras (cf. [44]).

dox algebraic logic.[6] In particular, the latter can nowise be qualified as "universal". It is rather our General Algebraic Logic that has really deserved to be referred to as really *Universal Algebraic Logic*.

As a matter of fact, the present book together with [42], [43] and [44] constitute *advanced* foundations of General Algebraic Logic. And we are going to advance further this novel branch of modern Algebraic Logic elsewhere. We refrain from specifying this point here to avoid redundant appetizing the reader.

---

[6]It is then not surprising that certain quite dishonest kleptomaniacs in orthodox algebraic logic have been inclined to plagiarize somebody else's work, in particular, Theorem 4.2 of [28], about which they did learn in 1995.

Bibliography

[1] F. G. Asenjo and J. Tamburino. Logic of antinomies. *Notre Dame Journal of Formal Logic*, 16:272–278, 1975.

[2] R. Balbes and P. Dwinger. *Distributive Lattices*. University of Missouri Press, Columbia (Missouri), 1974.

[3] N. D. Belnap, Jr. A useful four-valued logic. In J. M. Dunn and G. Epstein, editors, *Modern uses of multiple-valued logic*, pages 8–37. D. Reidel Publishing Company, Dordrecht, 1977.

[4] S. Burris and H. P. Sankappanavar. *A Course in Universal Algebra*. Springer-Verlag, New York, 1981.

[5] P. M. Cohn. *Universal Algebra*. D. Reidel Publishing Company, Dordrecht, 2nd edition, 1981.

[6] M. Dummett. A propositional calculus with denumerable matrix. *Journal of Symbolic Logic*, 24:97–106, 1959.

[7] J. M. Dunn. Algebraic completeness results for R-mingle and its extensions. *Journal of Symbolic Logic*, 35:1–13, 1970.

[8] E. Fried, G. Grätzer, and R. Quackenbush. Uniform congruence schemes. *Algebra Universalis*, 10:176–189, 1980.

[9] G. Gentzen. Untersuchungen über das logische Schliessen. *Mathematische Zeitschrift*, 39:176–210, 405–431, 1934.

[10] M. L. Ginsberg. Multivalued logics: a uniform approach to reasoning in artificial intelligence. *Computational Intelligence*, 4:265–316, 1988.

[11] G. Grätzer. *Lattice Theory. First concepts and distributive lattices*. W. H. Freeman and Company, San Francisco, 1971.

[12] G. Grätzer. *General Lattice Theory*. Akademie-Verlag, Berlin, 1978.

[13] G. Grätzer. *Universal Algebra*. Springer-Verlag, Berlin, 2nd edition, 1979.

[14] G. Grätzer and E.T. Schmidt. Ideals and congruence relations in lattices. *Acta. Math. Acad. Sci. Hungar.*, 9:137–175, 1958.

[15] K. Hałkowska and A. Zajac. O pewnym, trójwartościowym systemie rachunku zdań. *Acta Universitatis Wratislaviensis. Prace Filozoficzne*, 57:41–49, 1988.

[16] J. A. Kalman. Lattices with involution. *Transactions of the American Mathematical Society*, 87:485–491, 1958.

[17] H. Lakser. Principal congruences of pdeudocomplemented distributive lattices. *Proceedings of the American Mathematical Society*, 37:32–37, 1973.

[18] J. Łukasiewicz. O logice trójwartościowej. *Ruch Filozoficzny*, 5:170–171, 1920.

[19] A. I. Mal'cev. *Algebraic systems*. Springer Verlag, New York, 1965.

[20] E. Mendelson. *Introduction to Mathematical logic*. D. Van Nostrand Company, New York, 2nd edition, 1979.

[21] G. C. Moisil. Recherches sur l'algèbre de la logique. *Annales Scientifiques de l'Université de Jassy*, 22:1–117, 1935.

[22] C.S. Peirce. On the Algebra of Logic: A Contribution to the Philosophy of Notation. *American Journal of Mathematics*, 7:180–202, 1885.

[23] V. M. Popov. Sequential formulations of some paraconsistent logical systems. In V. A. Smirnov, editor, *Syntactic and semantic investigations of non-extensional logics*, pages 285–289. Nauka, Moscow, 1989. In Russian.

[24] A. P. Pynko. A structural semantic approach to constructing propositional logical systems. Preprint 1815, Space Research Institute of Russian Academy of Sciences, Moscow, February 1992. In Russian.

[25] A. P. Pynko. Algebraic study of Sette's maximal paraconsistent logic. *Studia Logica*, 54(1):89–128, 1995.

[26] A. P. Pynko. Characterizing Belnap's logic via De Morgan's laws. *Mathematical Logic Quarterly*, 41(4):442–454, 1995.

[27] A. P. Pynko. Definitional equivalence and algebraizability of generalized logical systems. *Annals of Pure and Applied Logic*, 98:1–68, 1999.

[28] A. P. Pynko. Functional completeness and axiomatizability within Belnap's four-valued logic and its expansions. *Journal of Applied Non-Classical Logics*, 9(1/2):61–105, 1999. Special Issue on Multi-Valued Logics.

[29] A. P. Pynko. Implication systems for many-dimensional logics. *Reports on Mathematical Logic*, 33:11–27, 1999.

[30] A. P. Pynko. Implicational classes of De Morgan lattices. *Discrete mathematics*, 205:171–181, 1999.

[31] A. P. Pynko. Regular bilattices. *Journal of Applied Non-Classical Logics*, 10(1):93–111, 2000.

[32] A. P. Pynko. Fuzzy semantics for multiple-conclusion sequential calculi with structural rules. *Fuzzy Sets and Systems*, 121(3):27–37, 2001.

[33] A. P. Pynko. Implicational classes of De Morgan Boolean algebras. *Discrete mathematics*, 232:59–66, 2001.

[34] A. P. Pynko. The completeness of derivable rules of labeled calculi for finite-valued logics. *Bulletin of Symbolic Logic*, 8(1):175–176, 2002. Abstract.

[35] A. P. Pynko. Extensions of Hałkowska-Zajac's three-valued paraconsistent logic. *Archive for Mathematical Logic*, 41:299–307, 2002.

[36] A. P. Pynko. A cut-free Gentzen calculus with subformula property for first-degree entailments in LC^*. *Bulletin of the Section of Logic*, 32(3): 137–146, 2003.

[37] A. P. Pynko. Semantics of multiplicative propositional signed sequent calculi with structural rules. *Journal of Multiple-Valued Logic and Soft Computing*, 10:339–362, 2004.

[38] A. P. Pynko. Sequential calculi for many-valued logics with equality determinant. *Bulletin of the Section of Logic*, 33(1):23–32, 2004.

[39] A. P. Pynko. A relative interpolation theorem for infinitary universal Horn logic and its applications. *Archive for Mathematical Logic*, 45:267–305, 2006.

[40] A. P. Pynko. Many-place sequent calculi for finitely-valued logics. *Logica Universalis*, 4(1):41–66, 2010.

[41] A. P. Pynko. Subquasivarieties of implicative locally-finite quasivarieties. *Mathematical Logic Quarterly*, 56(6):643–658, 2010.

[42] A. P. Pynko. *Equivalent universal Horn theories. General Algebraic Logic*. GRIN Verlag, Munich, 2018. ISBN 978-366-8-86946-2. URL https://www.grin.com/document/452732.

[43] A. P. Pynko. *Abstract sequent axiomatizations of finitary universal Horn theories. Abstract Proof Theory versus General Algebraic Logic*. Kindle Direct Publishing, 2019. ISBN 978-179-4-59554-5.

[44] A. P. Pynko. *Equivalential universal Horn theories. Algebraic systems versus pure algebras within General Algebraic Logic*. Kindle Direct Publishing, 2019. ISBN 978-179-4-51755-4.

[45] H. Rasiowa. *An Algebraic Approach to Non-Classical Logics*. North-Holland Publishing Company, Amsterdam, 1974.

[46] H.P. Sankappanavar. A characterization of principal congruences of De Morgan algebras and its application. In A.I. Arruda, R. Chuaqui, and N.C.A. da Costa, editors, *Proceedings of IV Latin American Symposium on Mathematical Logic, Santiago, 1978*, pages 341–349, Amsterdam, 1980. North-Holland Publishing Company.

[47] A. M. Sette. On the propositional calculus P^1. *Mathematica Japonica*, 18:173–180, 1973.

[48] M. Tokarz. Deduction theorems for RM and its extensions. *Bulletin of the Section of Logic*, 6(2):67–69, 1977. Abstract.

[49] M. Tokarz. Deduction theorems for RM and its extensions. *Studia Logica*, 38(2):105–111, 1979.

[50] A. Zbrzezny. The Hilbert-type axiomatization of some three-valued propositional logic. *Zeitschrift für Mathematische Logik und Grundlagen der Mathematik*, 36:415–421, 1990.

Appendix A

Sequent calculi for finitely-valued logics with equality determinant

During this appendix, $\alpha, \beta \in \{\omega, \omega \setminus 1\}$.

Fix a sentential language F and a finite non-one-element F-matrix $\mathcal{A}$ with finite *equality determinant* Υ (cf. [38]), that is, $x_0 \in \Upsilon \subseteq \mathrm{Tm}_F(V^1)$, while $\mathcal{A} \models ((x_0 \approx x_1) \leftrightarrow \bigwedge_{\phi \in \Upsilon}(D(\phi) \leftrightarrow D(\phi[x_0/x_1])))$. Set $\varepsilon_\Upsilon \triangleq \{\phi[x_0/x_i] \mapsto (\phi[x_0/x_{1-i}]) | \phi \in \Upsilon, i \in 2\} \subseteq \mathrm{Seq}_F^{[\alpha,\beta]}$, for $1 \in \alpha \cap \beta$. It is naturally identified with the equally denoted parameter-less purely relational $(F + \approx, F^{[\alpha,\beta]})$-translation given by $\varepsilon_\Upsilon(\approx) \triangleq \varepsilon_\Upsilon$. Then, we have (cf. (C.1) of [43]):

$$(\text{A.1}) \qquad (\mathfrak{A} + \approx) = (\varepsilon_\Upsilon)^{-1}[\mathcal{A}^{[\alpha,\beta]}].$$

Let $S_0^{\mathcal{A}} \triangleq A \setminus D^{\mathcal{A}}$ and $S_1^{\mathcal{A}} \triangleq D^{\mathcal{A}}$. For any $i \in 2$, put $\gamma_i^{\mathcal{A}} \triangleq \max\{|B| \, | \, \mathfrak{B} \in \mathbf{S}(\mathfrak{A}), B \subseteq S_i^{\mathcal{A}}\} \subseteq 2$ (cf. Lemma 7 of [39]). The work [38] suggests an enlargable F-sequent calculus $\mathbb{G}$ of type $[\omega \setminus \gamma_0^{\mathcal{A}}, \omega \setminus \gamma_1^{\mathcal{A}}]$ with structural rules, said to be *associated with* $\mathcal{A}$, such that:

$$(\text{A.2}) \qquad \vdash_{\mathbb{G}} = \vdash_{\mathcal{A}^{[\omega \setminus \gamma_0^{\mathcal{A}}, \omega \setminus \gamma_1^{\mathcal{A}}]}}.$$

Made in the USA
Monee, IL
07 July 2026

56551418R00039